Kea, Bird of Paradox

A fledgling kea plays with a branch at the study site in Arthur's Pass National Park, on the South Island of New Zealand. (Photograph by Judy Diamond)

Kea, Bird of Paradox

The Evolution and Behavior of a New Zealand Parrot

Judy Diamond and Alan B. Bond

UNIVERSITY OF CALIFORNIA PRESS

Berkeley   Los Angeles   London

University of California Press
Berkeley and Los Angeles, California

University of California Press, Ltd.
London, England

© 1999 by
The Regents of the University of California

Library of Congress Cataloging-in-Publication Data

Diamond, Judy.
  Kea, bird of paradox : the evolution and behavior of a New Zealand
parrot / Judy Diamond and Alan B. Bond.
      p. cm.
  Includes bibliographical references and index.
  ISBN 0-520-21339-4 (alk. paper)
  1. Kea—Evolution. 2. Kea—Behavior. I. Bond, Alan B., 1946– . II. Title.
QL696.P7D535 1999
598.7′1—dc21   98-19180
                                                                             CIP

The epigraphs are from the following sources: Philip Temple, *The Legend of the Kea, or How Kritka Stole the Best Beak and Best Claws from Ka, the Great Bird of All Birds, and Took the Keas to Live in the Highest Mountains* (Auckland: Hodder and Stoughton, 1986); used by permission of the author. Jared Diamond, "Bob Dylan and Moas' Ghosts," *Natural History* 99, no. 10 (1990): 31; used with permission of *Natural History*, copyright the American Museum of Natural History 1990. Alfred Russel Wallace, *Darwinism* (London: MacMillan, 1891), 75. J. R. Jackson, "Life of the Kea," *Canterbury Mountaineer* 31 (1962): 122. Niko Tinbergen, "On Aims and Methods of Ethology," *Zeitschrift für Tierpsychologie* 20 (1963): 420; used with permission of Blackwell Science Ltd. Niko Tinbergen, "The Evolution of Behavior in Gulls," *Scientific American* 203, no. 6 (1960): 130. Kerry-Jayne Wilson, "Kea: Creature of Curiosity," *Forest and Bird* 21 (1990): 26; used with permission of *Forest and Bird*.

Printed in the United States of America
9 8 7 6 5 4 3 2 1
The paper used in this publication meets the minimum requirements of American National Standards for Information Sciences—Permanence of Paper for Printed Library Materials, ANSI Z39.48-1984.

To Benjamin and Rachel

CONTENTS

List of Illustrations ix

Acknowledgments xi

Introduction 1

ONE The Moa's Legacy
7

TWO From Relict to Renegade
26

THREE Hanging Out with the Gang
46

FOUR Growing and Learning
82

FIVE The Prince and the Pauper
101

SIX From Bounties to Black Markets
123

Appendix A: List of Common and Scientific Names 151

Appendix B: Supplementary Tables 156

Notes 175

References 201

Index 223

ILLUSTRATIONS

Fledgling kea playing with a branch	*frontispiece*
Comparison of moa and ostrich skeletons	9
Haast's eagle attacking a moa	13
Two keas on a mountain beech	15
Evolution of kea and kaka	23
Kea attacking a sheep	33
Keulemans's lithograph of the kea	37
Comparison of fledgling and adult keas	52–53
Measuring a kea's upper bill	54
Comparison of male and female kea bills	55
Kea facial expressions	61
Subadult male kea allopreening a juvenile male	65
Adult male kea feeding a fledgling	67
Pair of fledgling male keas jumping and flapping	71
Pair of juvenile male keas during tussle play	72
Three juvenile keas playing tug-of-war	79
Four-day-old kea nestling	83

Fledgling and juvenile keas demolishing a chair	87
The wing-hold display	88
The hunch display	90
North Island kaka on Kapiti Island	103
Comparison of kea and kaka bills	104
Female kaka in pout display	113
Adult male kaka feeding a nesting female	113
Kea feeding on sheep carcass	132
Keas destroying an automobile	137
A car protected from keas	138
Publicity for the "Do Not Feed the Kea" campaign	139

MAPS

New Zealand	3
The South Island	29
Distribution of keas in New Zealand	144

TEXT TABLES

1.	Common kea foods	17
2.	New Zealand birds extinct in the last thousand years	43–45

ACKNOWLEDGMENTS

No scientific undertaking of this size is possible without substantial assistance from various individuals and organizations. The National Geographic Society funded our research, including all travel costs for three field seasons in Arthur's Pass National Park, New Zealand. We received additional support from the University of Nebraska State Museum.

Many people in New Zealand facilitated our research. We are especially grateful to Peter Simpson and Ron Moorhouse of the New Zealand Department of Conservation. As Senior Conservation Officer at Arthur's Pass National Park, Peter provided us with extensive documentation on climate, kea mortality, and government policies, as well as valuable anecdotes about keas. He also offered all manner of logistical support, including securing housing for us in park facilities. Ron served as our host, tour guide, and housemate on Kapiti Island; he gave us full details of his thesis research on kakas as well as data on his banded population and also loaned us a hide for observing the birds.

Peter Daniel, also of the Department of Conservation, transported us back and forth from the mainland to Kapiti Island and hospitably housed us in the old whare. He further allowed us to control the feeding times of the kakas, enabling us to maximize our data.

Kerry-Jayne Wilson of Lincoln University, Canterbury, graciously allowed us to trap and band keas under her permit and loaned us equipment to do so. She and her student Ria Brejaart also supplied data on measurements and sightings of banded keas in the Arthur's Pass and Craigieburn areas. Their data proved invaluable in determining the size and range of movements of our local population.

We received significant help and information from many staff members of the Department of Conservation, both at Arthur's Pass National Park and elsewhere in New Zealand. Particular thanks go to Steve Phillipson, Andy Grant, and Robin Smith. We also thank Jacqueline Beggs for giving us a tour of her kaka research operation in Nelson Lakes National Park, and David Falwell of the Auckland Zoo and Paul Barnett of the Wellington Zoo for providing tours of their facilities and firsthand accounts of the behavior of keas and kakas in captivity. The residents of Arthur's Pass Village showed us every kindness and hospitality and were unusually tolerant of our intrusion on the privacy of their refuse dump.

In the United States, the staff of the San Diego Zoo granted us entry to many of their facilities so that we could test our recording techniques and observational methods. We especially thank Alan Lieberman and Wayne Schulenburg for giving us access to their captive kea population.

A number of museums around the world loaned us material for our study of kea sexual dimorphism. We are most grateful to Amadeo Rea, Curator of Birds and Mammals, who allowed us to use his facilities at the San Diego Natural History Museum. We also thank P. R. Millener of the National Museum at Wellington, Brian Gill of the Auckland Institute and Museum, and G. A. Tunnicliffe of the Canterbury Museum in Christchurch for access to their collections.

Finally, we would like to thank Anne Wertheim Rosenfeld, Albert Baez, and the late Dr. Bernard L. Diamond for their help at critical points in this project.

INTRODUCTION

"But where is the kea's place, Ka? Where must we go to perch?" Ka's voice boomed with thunder. "You are too late, foolish bird. All the perches are given out."

"Then what shall we do?" cried Kritka.

"You have the best claws and the best beak, Kritka," said Ka. "You need no special place to perch." Ka's laughter crackled through the sky and his feathers dazzled the world as they reflected the rays of the sun.

Kritka flew after Ka. He circled higher and higher and tried to catch the bird of all birds before he disappeared into his own darkness. Kritka flew so high that the setting sun shone on his underwings. He captured its red colour for the feathers of all other keas.

Philip Temple, *The Legend of the Kea*

In the rugged mountains of New Zealand lives a crow-sized parrot called the kea. This bird possesses an extraordinary, alien intelligence that has given it a paradoxical reputation as both a playful comic and a vicious killer. Its true character is a mystery that biologists around the globe have debated for more than a century. Based on our four-year field study, this book presents the first comprehensive account of the kea's contradictory nature.

The kea evolved in one of the most isolated parts of the world. The New Zealand archipelago, located in the southwestern Pacific Ocean, consists of three primary land masses—the North Island, the South Island, and Stewart Island—and over five hundred smaller islands scattered over twenty degrees of latitude, from the subtropical Kermadecs in the north to cold Campbell Island in the south, only 1,800 km from Antarctica. Unlike most other oceanic islands, however, New Zealand is geologically ancient, a fragment of a primeval continent that has been separated for 80 million years from the rest of the world by huge expanses of ocean.

In such relative isolation, evolution took new directions, resulting in a distinctive and sometimes bizarre assemblage of plants and animals that are wholly unlike anything found elsewhere. Jared Diamond, the distinguished evolutionary ecologist, has remarked that "New Zealand is as close as we will get to the opportunity to study life on another planet."[1] The land mammals that dominate the ecology of other continents were absent from New Zealand until the arrival of humans. Instead, a unique collection of birds evolved, including giant moas, kiwis, flightless parrots and rails, and tiny flightless wrens. But the most extraordinary animal in this extraordinary fauna is the kea, the world's only alpine parrot.

Keas are native only to the South Island, breeding mainly in the Southern Alps, a mountain range that forms the island's rocky spine. A large portion of the South Island lies above tree line, and hundreds of its peaks are covered with perennial snow.

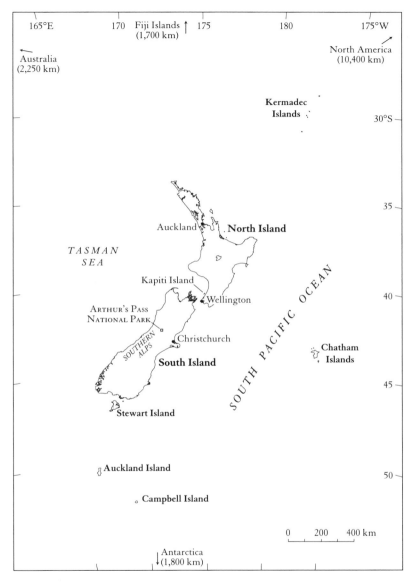

New Zealand. (Map by Bill Nelson)

In this rigorous and unforgiving environment the kea has evolved a level of intelligence and flexibility that rivals that of the most sophisticated monkeys. These birds are bold and curious, infuriatingly persistent, and ingeniously destructive. They display complex social behavior with intricate rules to determine priority in feeding. And they show more elaborate and extensive play behavior than any other bird.

In the nineteenth century, naturalists believed keas to be fruit and nectar feeders, like many other native birds. When sheep were introduced into New Zealand, however, keas were soon hunted as outlaws, accused of attacking and tearing the flesh from defenseless livestock. Naturalists were amazed by the kea's apparent transformation within only a few decades. In fact keas may have originally fed on carrion, as well as fruit, nectar, and insects. If so, their predation on livestock is less a transformation than a demonstration of their ability to adapt to changing conditions. This flexibility is the hallmark of what has been called an "open program" species—one that specializes in learning and in applying its skills in new ways to new circumstances.

Enormously resilient, keas feed on nearly any accessible resource and make easy use of refuse dumps, campsites, and other human environments. Yet their future existence is precarious, as conflicts with humans, habitat destruction, and poaching for the pet trade threaten the relatively small population that remains. The natural history of the kea thus demonstrates the limitations as well as the strengths of behavioral flexibility and provides lessons applicable to other species, including our own.

To trace the evolution of the kea's behavior, we reconstruct the role of the kea in New Zealand's prehistoric ecology. Chapter 1 shows how events during the ice ages, when the geography of the islands was extremely unstable, may offer clues to the origin of the kea's personality. In chapter 2 we discuss the successive waves of human settlement, beginning a thousand years ago, that eliminated much of the original fauna. We see how the kea's extraordinary ability to cope with such environmental changes first brought this species to the attention of scientists.

Next we describe and interpret the present-day behavior of the kea from three perspectives. In chapter 3, we follow a group of keas through the course of a day and offer the first detailed descriptions of kea play in the wild. In chapter 4, we describe the developmental changes that take place over the birds' lifetime, as they grow from reckless and comic fledglings to sometimes cruel and aggressive adults. In chapter 5, we contrast the kea's rambunctious behavior with the more sedate nature of its closest relative, the kaka. These three perspectives reveal what keas learn as they grow to maturity, why they play so much more than other birds, and how flexibility as a lifestyle has become an evolutionary necessity in this species. Finally, chapter 6 considers the present-day interactions between keas and humans, the dangers posed to the survival of the species, and the struggle to achieve peaceful coexistence.

We use common names for all species in the text; the corresponding scientific names may be found in Appendix A. The results of the quantitative analyses that form the basis of our discussion of kea behavior are included as tables in Appendix B.

ONE

The Moa's Legacy

> New Zealand's moas are no longer alive, but they molded the anatomy and behavior of New Zealand's surviving species. Their ghosts still hang over the New Zealand landscape.
> Jared Diamond, "Bob Dylan and Moas' Ghosts"

The present-day flora and fauna of New Zealand bear little resemblance to those that made up the community in which keas originally evolved. Nearly half of all the species that thrived in New Zealand a thousand years ago vanished in one of the largest waves of extinctions in history. This "biological holocaust" began with the first arrival of humans on the islands and has continued up until recent times. A new set of actors has now taken over the ecological stage, filling roles that had been left vacant by the original occupants. To understand the biology of the kea, and the factors that influenced its evolution, we must begin by reconstructing the prehuman environment of the islands and repopulating it with the full range of original inhabitants. We must, in short, tell a ghost story.[1]

WHEN THE MOAS REIGNED

Our story begins on the South Island about eight thousand years ago, when New Zealand first emerged in its present form. Temperate rainforests extended in all directions in a virtually unbroken canopy, occupying the western and eastern slopes of the Southern Alps up to about 1,200 m elevation.[2] The highest tier consisted of *Nothofagus*, the southern beech, either in uniform stands or mixed with conifers and broad-leaved trees. Below were forests of podocarps, a group of conifers peculiar to the Southern Hemisphere, along with tree ferns and many different kinds of broad-leaved trees. In the northern parts of the South Island grew great forests of kauri, the largest and oldest trees in New Zealand. Only in the drier areas along the eastern plains and above tree line in the Alps did the forest give way to grassland and scrub.

This pristine wilderness had been isolated from other continents since the time of the dinosaurs. Instead of cattle, deer, and rabbits, New Zealand's grazers and browsers were all large, flightless birds. Gigantic moas, abundant on the South Island, were the most common browsing animal in the archipelago. There were eleven species of moas in six genera; one species included perhaps the tallest birds that ever lived. Similar in appearance to ostriches or emus, moas were adapted to a range of different forest and grassland habitats.[3] Like bison in the American West, they were the keystone of the ecosystem, shaping the flora and the landscape of New Zealand and dominating all other groups by their weight and numbers.

In the nineteenth century, when moa skeletons were first described, the birds were thought to have been enormous—by some accounts the size of a giraffe. This turn-of-the-century drawing compares the skeleton of the largest species of moa, *Dinornis giganteus (right)*, with the skeleton of an ostrich. On the left is Sir Richard Owen, the English scientist who first classified the moa in 1839. Although still considered the tallest bird that ever lived, *D. giganteus* is now thought to have been only slightly taller, but much heavier, than an ostrich. (Illustration from Pettigrew 1908, 1307)

Moas were found throughout the South Island, including the mountain forests, a habitat they must have shared with keas. *Megalapteryx didinus,* the most common moa on the South Island, was probably a mountain specialist, because the largest accumulations of its bones are from subalpine sites. Bones of larger moas, such as *Pachyornis* and *Dinornis,* have also been found in subalpine deposits. *Dinornis,* which were about the size of an elk, probably browsed on twigs, leaves, and fruit, occasionally taking fallen fruit from the forest floor. Smaller moas like *Megalapteryx,* which were about the size of an adult deer, may have been more omnivorous, including large invertebrates and insects in their diet. Moa chicks of all species may have fed mainly on insects and other small invertebrates. An analysis of moa gizzards suggests that moas ate at least some of the same plant foods as keas. For example, *Dinornis* seems to have fed heavily on the bright orange coprosma berries that are a staple in the kea's diet.[4]

Among the most common grazers in the moa's kingdom were takahes and kakapos, birds that are extremely rare today. Takahes are large flightless gallinules with heavy, shearing bills that give them the appearance of small blue dinosaurs. They dig up grasses and then hold them in one foot while delicately peeling away the fibrous outer layers. The lumbering kakapo, a flightless nocturnal parrot the size of a turkey, was one of the most abundant birds on the South Island in the moa's day. Kakapos dig up roots and bulbs, peel bark and foliage from trees, and nip off buds, fruits, and shoots of low-growing shrubs. They also strip the seeds from grass and chew the blades, which are left dangling in a characteristic wad from the leaf end.[5]

Many other, smaller, animals filled niches in the kingdom of the moas. Wetas, which are giant forest crickets, were then extremely common. They behaved much like mice, feeding on wood and leaves at night and leaping long distances when startled. They kept company in the underbrush with a diverse group of tiny, flighted and flightless New Zealand wrens, members of a unique family. A variety of smaller insects and worms, frogs, skinks, geckos, and giant land snails were common on the forest floor. These creatures probably served as food for tuataras, lizard-like reptiles the size of an iguana. Tuataras are the last remaining representatives of an ancient order of reptiles; they possess, among other odd features, a third, vestigial eye in the middle of their foreheads.[6]

A large array of birds fed on insects in addition to fruit and nectar. Many of these species belonged to bird families with no clear relatives anywhere else in the world. The New Zealand wattlebirds included the oriole-like saddlebacks and the large, gray kokakos. The most spectacular wattlebird, however, was the huia, whose bill size differed between the sexes to a greater degree than that of any other bird species: the female's bill was nearly one and a half times as long as the male's. Wattlebirds fed mainly in the trees, while piopios, another species of exclusively New Zealand birds, searched for food on the ground. The underbrush was also home to nocturnal kiwis, with their long, probing bills; wekas, which are hen-sized flightless rails; and owlet-nightjars, which were distant flightless relatives of the whippoorwill and nighthawk.[7]

In the daytime small insectivorous birds foraged in the foliage, among them warblerlike birds, such as whiteheads and yellowheads, and more robust species similar to chickadees, such as New Zealand robins and fantails. The nectar in flowering trees attracted stitchbirds, bellbirds, and tuis, members of a unique group of honey-eaters. Also fond of nectar are New Zealand parrots, particularly the kakarikis, which are New Zealand parakeets, and the kaka, the forest cousin of the kea; both go back to the time of the moas. One of the most beautiful birds in the moa's forest was the New Zealand pigeon, with its striking iridescent green and white plumage. New Zealand pigeons are consummate frugivores, putting away large volumes of fruits and leaves.[8]

There were many carnivorous birds as well. Some were nocturnal predators, such as laughing owls and moreporks, both fairly small owls that mostly preyed on insects. Moreporks, however, do apparently eat nestlings of other birds, and perhaps the now-extinct laughing owls did so too. Flightless adzebills, about the size of a turkey, could easily have killed and eaten large lizards, as well as newly fledged chicks and eggs of other birds.[9] But the main danger to adult birds would have been the larger raptors, which attacked from the air during daylight hours. Among them was one of the largest and most powerful birds of prey that ever lived, the Haast's eagle. Roughly the size of an Andean condor, it had a wingspan of nearly three meters. It may have been a "sit and wait" predator, observing from its high perch for long periods and then rapidly diving on its prey. Haast's eagles were certainly large enough to have fed on keas and may have been the sole predator capable of attacking the largest moas, whose carcasses would be consumed over several days.[10]

Medium-sized aerial predators included the New Zealand goshawk and the New Zealand falcon, which would chase down other birds in flight. The now-extinct goshawk weighed about two to three times as much as a kea, while the falcon is comparable to the kea in size. These predators probably swooped from high branches and swerved through the forest in pursuit of their prey. The goshawk was surely a savage threat to a variety of midsized birds, including keas, kakas, kakapos, New Zealand pigeons, wekas, and New Zealand ravens. The falcon generally preys on smaller birds, though it will attack larger ones when the opportunity arises.[11]

The presence of large predators opened a niche for scavengers. Today keas, harriers, and black-backed gulls all gather at car-

Haast's eagle was probably the largest eagle that ever lived. It has been extinct for about a thousand years. This huge predator most likely killed moas, whose carcasses would have been fed on by a community of scavengers, including keas. (Illustration by Colin Edgerley, from R. N. Holdaway 1989b; used by permission of *New Zealand Geographic*)

casses and strip the remaining meat from the bones. In the moa's kingdom, however, the main scavenger was most likely the New Zealand raven. These birds were larger than keas and were abundant both in forest and in adjacent scrub regions. They were probably at least nominal omnivores, feeding on fruit as well as large insects, lizards, tuataras, and small birds. They undoubtedly scavenged carrion from moas killed by Haast's eagles and possibly preyed on the eggs and young of medium-sized birds, such as keas and kakas. They may even have fed on moa eggs and chicks. Raven and kea fossils have been found together in cave deposits, so the ranges of the two species overlapped, and both may have scavenged at the same carcasses.[12]

THE KEA'S NICHE

At the time of the moas, keas probably fed much as they do today. Beech trees provide the bulk of their diet. Keas are more closely associated with forests of southern beech than with any other habitat. Their olive-green plumage blends perfectly with the somber shades of the beech foliage. The birds roost in the trees, nest in burrows in the depths of beech forest, and take shelter among the thick branches against predators or inclement weather. They also feed in the canopy and the second tier of beeches on whatever is seasonally available: buds, leaves, or nuts. When beechnuts are abundant, they forage for hours in the crowns of the trees. Keas pick the nuts off one by one, using their bills like a forceps, and then grind and crush the nuts between their lower bills and their hard palates.

The resources of the beech forests, however, are not enough to sustain keas. The low diversity in the understory of shrubs

Keas are closely associated with forests of mountain beech, which blanket many of the high alpine areas on the South Island. (Illustration by Mark Marcuson)

and forbs limits the amount of fruit, nectar, and edible foliage. Furthermore, beeches, like many other New Zealand plants, are mast-seeding, which means that they produce enormous volumes of nuts for one or two seasons and then very little for many years. From one year to the next, beech can be a very unreliable resource.[13]

Keas must therefore seek out whatever food is seasonally abundant. In springtime they dig up large mountain daisies in the alpine grasslands, sometimes consuming the entire plant, roots and all. They also search at the edges of snow mounds, probing and digging around rocks for low-growing plants and insects. In summer keas forage in the alpine shrub habitat for fruit, foliage, seeds, and flowers. The birds relish the orange berries of coprosma bushes, of which there are over forty-five species. After beech trees, coprosma berries are probably their most important food source, but they also consume the red berries of the mountain totara. Keas feed readily from flowers, clinging to the branches of rata trees or mountain flax as they rapidly lap up the nectar and pollen. They also catch and eat huge numbers of grasshoppers, beetle grubs, and other insects.[14]

In autumn keas spend much of their time feeding in the forest, where mountain beech buds and young leaves are plentiful. As in summer, they also forage on mountain daisy roots and coprosma berries. Roots, bulbs, stems, fruit, and seeds all continue to be important sources of food. The winter months of June to September, however, are the time of greatest mortality for keas, mainly due to starvation. The birds often feed below tree line on the forest floor, scrounging for remnants of fall berries. They also avidly seek animal fat and will fight vigorously to obtain it. They tear open carcasses to consume meat and internal organs; they scrape dried meat from bones, which they then open at one end to lick out the marrow.[15]

IN THE COMPANY OF GHOSTS

Keas, then, are truly omnivorous, with a breadth of interest in plant and animal foods that roughly matches that of humans or coyotes in the American West. They consume an amazing array

TABLE 1.   Common Kea Foods

Food Source	References
PLANTS	
astelia	Riney et al. 1959, 40
beech, mountain	Riney et al. 1959, 40; Campbell 1976, 15; Clarke 1970; Brejaart 1988, 45
beech, silver	O'Donnell and Dilks 1986
coprosma	Riney et al. 1959, 39; Campbell 1976, 16, 23
daisy, mountain	Riney et al. 1959, 39; Brejaart 1988, 45; Myers 1924; Jackson 1960; Campbell 1976, 16
flax, mountain	Myers 1924
lily, mountain	Myers 1924
rata, southern	Jackson 1960; O'Donnell and Dilks 1986
snow groundsel	Riney et al. 1959, 39
spear-grass	Myers 1924
totara, mountain	Jackson 1960; Campbell 1976, 16
ANIMALS	
beetle grubs	Brejaart 1988, 42; Myers 1924; Clarke 1970; Campbell 1976, 24
deer, red	Riney et al. 1959, 40; White 1894; Benham 1906
grasshoppers	Brejaart 1988, 42; Myers 1924; Clarke 1970; Campbell 1976, 24
rabbits	Marriner 1908, 144
rats and mice	Porter 1947; Jackson 1962b; Yealland 1941
sheep	White 1894; Aspinall 1990, 6
snails, land	Meads, Walker, and Elliott 1984

of foodstuffs, including as many as a hundred species of plants and animals (common examples in table 1).[16] Very few of these resources, however, would have been exclusively theirs in the moa's kingdom. Keas had to compete with kakas for rainforest fruits and nectar, with kakapos and takahes for bulbs and succulent grasses, with moas for fruits and leaves, with the guild of insectivores for insects and grubs, and with ravens, harriers, and gulls for access to carrion. Only southern beech provided a staple that would have been relatively uncontested, perhaps because, apart from the rare years of high nut production, it was generally a marginal resource.[17]

For all other foodstuffs the competition was probably fierce; many of the competitors were undoubtedly far more common than keas. Moas are found in every type of subfossil deposit in New Zealand. Moreover, because of their great size, even one moa would have consumed an enormous volume of leaves and fruits. Among the smaller birds kakas were abundant, and their range overlapped with the kea's at all but the highest altitudes. Even today, under protected conditions, their populations can easily be twenty times as dense as those of keas. In the moa's kingdom, kakapos and takahes would possibly have rivaled—in tonnage, if not in numbers—the rabbits that have since occupied their niche in the New Zealand fauna.[18]

In a world of dietary specialists the kea survived as the ultimate generalist, feeding on almost anything that came its way and actively searching out alternative sources of food. Although kakas might open logs for beetle grubs more quickly, takahes might peel daisy bulbs more efficiently, and ravens might tear apart carcasses more vigorously, keas, when given the opportunity, could do all these things adequately.

With few exceptions, the species that once directly competed with keas or preyed on them have been eliminated from kea habitat or are either extinct or vastly reduced in numbers. The moas are gone, as are the eagles, goshawks, ravens, and adzebills. Kakas are restricted to remnant tracts of old-growth, native forest, and kakapos and takahes are now among the world's rarest birds. The kea lives on in a phantom community of predators, browsers, grazers, frugivores, and scavengers that once shaped its broad generalist strategy. It has survived in part because of just this ability to make use of whatever resources chance brings its way.

THE ORIGINS OF NEW ZEALAND

The kea is a product not only of its recent history but also of the conditions that made its evolution possible. These are rooted in the vast changes that occurred since New Zealand became a separate continent. In the age of dinosaurs, New Zealand was fused with the other southern continents into a single gigantic landmass, the supercontinent Gondwanaland. In this jigsaw puzzle of continents, New Zealand lay between the Pacific coast of western Antarctica and the east coast of Australia. Plants and animals dispersed throughout the region, moving easily through areas that today are separated by broad stretches of ocean. Forests of southern beech covered large parts of the supercontinent.[19]

As Gondwanaland broke up about 80 million years ago, its component parts drifted off across the Southern Hemisphere, carrying the forests and their occupants with them. These floating continents bore southern beeches and lowland conifers, such as kauri and rimu, and many ferns and mosses. Today forests of southern beech provide a reminder of a common geographical bond. Trees related to New Zealand's southern beeches can be found in Chile, Tasmania, New Guinea, and New Caledonia.[20]

Animals rode along with the trees. New Zealand's fauna contains many creatures that trace their ancestry to Gondwanaland forests, including frogs, skinks, geckos, land snails, some unusual spiders, and several groups of freshwater insects. The tuatara originated at this time, as did the wetas. The peripatus, an odd caterpillar-like animal that shares many of the attributes of arthropods and earthworms, is still found in New Zealand forests, as well as in South America and Africa, little changed from its Gondwanaland ancestors. The predecessors of the giant flightless moas may also have drifted along with the continents.[21]

In the early stages of the continental separation, while the oceans were expanding between New Zealand, Australia, and New Caledonia, many organisms took advantage of the archipelagos that still tenuously connected the regions, working their way across by island-hopping. By the end of the age of dinosaurs, 65 million years ago, most of these islands were gone, and a new ocean, the Tasman Sea, stood between New Zealand and Australia. In spite of the now formidable distances, species continued to arrive, either blown on storm winds or drifting on rafts of floating vegetation. Even today, nearly 80 percent of the genera of higher plants found in New Zealand are shared with Australia.[22]

As the continents became more isolated and their climates more distinctive, the flora and fauna of Australia and New Zealand began to diverge. Australia developed a diverse array of snakes and marsupial mammals that were apparently unable to cross the Tasman Sea. Aside from two species of bats, which must have arrived by air, New Zealand has no native land mammals and no reptiles other than lizards and tuataras. The absence of such large terrestrial predators was one of the major factors that shaped the strange and fragile ecology of the islands.[23]

THE EVOLUTION OF THE KEA

Between 50 million and 25 million years ago, while mammals were diversifying and taking over the rest of the world, New

Zealand was undergoing great tectonic upheavals. Mountains rose and fell, causing changes in sea level and severe fluctuations in climate. At the beginning of this period New Zealand was about 22 percent larger than it is today. By the end it had shrunk to only 18 percent of its present size. With this drastic reduction in land surface, species that had never before come in contact were forced to compete for limited resources, and a mass extinction of land organisms ensued.[24] As the land rose again and stabilized, the lost species were gradually replaced by new groups invading from surrounding regions.

The ancestor of the three species of parrot in the genus *Nestor*—the kea; its brown cousin, the kaka; and their close relative, the Norfolk Island kaka—probably came from Australia. But the taxonomic evidence of a relationship between *Nestor* and the Australian parrots and parakeets is not definitive. It would have to be a very distant kinship, indicating a long history of separate evolution. Thus, the ancestral *Nestor* may have arrived in New Zealand as much as 20 million years ago.[25] A forest-dwelling parrot, the "proto-kaka," is presumed to have lived in New Zealand about 15 million years ago, when the region consisted of a single, large island. At that point the geology of New Zealand remained comparatively quiet for about the next 10 million years.[26]

At the start of the Pleistocene, about 2 million years ago, this stability was shattered. New Zealand entered a tumultuous period of sharply oscillating climatic conditions. As temperatures cooled, glaciers flowed out from the mountains and covered the land. During these glacial periods the forests retreated and sea levels dropped, causing smaller islands to join the mainland. When warm temperatures returned, during the interglacial periods, the glaciers receded, the forests spread, and sea levels rose, inundating low-lying areas and dissecting New Zealand again into many small islands, which would later merge in colder periods. As a result of this fluctuation, the contrast between northern and southern environments became extreme. In the south, at the last glacial maximum, about 22,000 to 14,000 years ago, a continuous chain of glaciers and ice sheets stretched over the Southern Alps. In the north, however, temperatures were milder and glaciation was relatively minor.[27]

This geographic contrast, combined with the repeated fragmentation and reunification of the habitat, is a classic recipe for generating new species.[28] When a continuous, interbreeding population is separated by climatic or geographic barriers that create distinctive habitats, the populations are subjected to different selective pressures and may evolve different features. When the barriers are subsequently removed, the accumulated differences may be sufficient to prevent the populations from freely interbreeding again; in other words, they will have become separate species. The longer the populations remain isolated from each other and the more extreme the habitat differences they experience, the more likely it is that they will diverge and become new species.

The proto-kaka may have diverged into new species sometime in the early Pleistocene. Several populations of this forest parrot were physically separated from one another by the subdivision of the islands as sea levels rose. The population in the more benign north became kakas, specializing in exploiting the fruits and insects of the rainforest. The population living in the harsher southern regions eventually became keas, developing the behavioral strategies and food preferences that would help them survive among the ice fields.[29]

The Moa's Legacy / 23

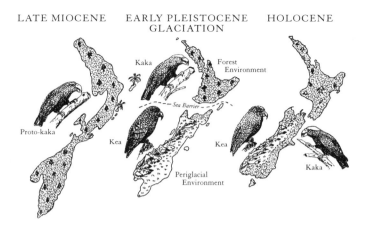

LATE MIOCENE   EARLY PLEISTOCENE   HOLOCENE
                    GLACIATION

About 15 million years ago the proto-kaka, a parrot of the genus *Nestor*, lived on the large single landmass that was New Zealand. When New Zealand separated into smaller islands during the early Pleistocene, two populations of proto-kakas were isolated. As the environments of the north and the south became more distinct, the two populations diverged. When the daughter stocks (kea and kaka) eventually were reunited, they coexisted as separate species. (Illustration by C. A. Fleming, from Fleming 1979; used by permission of Auckland University Press and M. A. Fleming)

When keas first evolved, forests were sparse on the South Island. Pollen records suggest that all trees except southern beeches were rare. Beneath the glaciers and snow much of the South Island was covered in grassland and open herb fields. Many of the alpine communities of the South Island developed during this period, producing a flora of remarkable diversity and uniqueness. Over half the native plant species in New Zealand are restricted to the alpine regions, and many of these are found nowhere else on earth.[30]

Evidence from fossils found in limestone caverns indicates

that keas were quite common in the mid-Pleistocene. Their flexibility and cleverness, products of their evolution amid the unforgiving ice and snow, stood them in good stead in exploiting the erratic, patchy food supply of the cold grasslands. As the ice retreated, between 14,000 and 10,000 years ago, grassland gradually gave way to denser shrub land in the South Island, but there was still relatively little forest cover. Then, about 9,500 years ago, pollen records show a dramatic change. In less than a thousand years, forests of not only beech but also conifers and broadleaved trees advanced to blanket most of the South Island.[31]

Coincidentally, about nine thousand years ago, something led to a severe reduction in the kea population. The fossil records suggest that at about the time when the forests of the South Island were expanding, keas were becoming rare. What happened is far from clear. The increase of the forests should not, in itself, have threatened the kea. Although the birds are primarily adapted to alpine conditions, they readily make use of the abundant resources available in lowland forests. In some fossil caves keas continued as a major component of the fauna even after the shift from subalpine and montane to lowland forest conditions. Thus, habitat modification cannot have been solely responsible for the species' subsequent abrupt decline.[32]

Clues to the mystery may lie instead in the company of competitors that the kea came to face in a milder and more hospitable climate. When the forests returned, they presumably brought with them the characteristic fauna of the moa's kingdom, including the forest-adapted kaka. Confronted by these foraging specialists, keas would have been forced back into a receding alpine habitat, reduced to a sparse and less rewarding spectrum of foods. As a result, their numbers dwindled.[33] But even so, their shrewdness and resiliency allowed some of them to survive, making their living in the hostile alpine environment or in the neglected corners and interstices of the forest, where they would scrounge bits that the specialists had overlooked or not yet gotten around to. An uncommon species of harsh and marginal habitats, keas might have continued indefinitely in this fashion had it not been for the arrival of humans in New Zealand.

TWO

From Relict to Renegade

> About 1868 [the kea] was first observed to attack living sheep, which had frequently been found with raw and bleeding wounds on their backs. Since then it is stated that the bird actually burrows into the living sheep, eating its way down to the kidneys, which form its special delicacy. As a natural consequence, the bird is being destroyed as rapidly as possible, and one of the rare and curious members of the New Zealand fauna will no doubt shortly cease to exist. The case affords a remarkable instance of how the climbing feet and powerful hooked beak developed for one set of purposes can be applied to another altogether different purpose, and it also shows how little real stability there may be in what appears to us the most fixed habits of life.
> Alfred Russel Wallace, *Darwinism*

FIRST WAVE: MAORI SETTLEMENT

Human habitation of New Zealand began only about a thousand years ago, when a seafaring people from Polynesia first settled on the east coast of the South Island. These earliest settlers clung mainly to the coastline, fishing and hunting the abundant wildlife and traveling to new areas by canoe. They subsequently dispersed north and south to occupy territory throughout the three main islands. Their descendants, the Maori, eventually became concentrated on the North Island, with fewer settlements on the South Island. By the 1600s the population of the South Island had stabilized at about ten thousand, most of whom were members of the Ngai Tahu clan.[1]

The Maori people coined new words for the strange collection of plants and animals that made up the islands' ecology. They named the green parrot of the high mountains "kea," after the bird's exultant cry, and gave the lowland brown parrot the name "kaka," possibly after one of its characteristic vocalizations. ("Kaka" is also used as a general term for "parrot" throughout Polynesia, where it means "to chatter or gossip.")[2]

The kaka and the New Zealand pigeon were among the most common forest game birds hunted by the Maori. A Maori saying likens the kaka to a prevalent game fish: "What the barracouta is at sea, so the kaka is on land." Kakas were taken by hand, with spears, with decoy birds and either slip nooses or more elaborate snares, or with a striking rod when the birds were drinking. They were in many respects a central element of Maori culture. Their bones and beaks were fashioned into tools and decorations; their feathers were sewn into elaborate ceremonial cloaks. The Maori referred to them with an extensive, specialized vocabulary.[3]

Keas, however, do not seem to have played a similar role. They receive scant mention in Maori folklore, in which birds are central actors. A member of the Ngai Tahu, interviewed on the South Island during the 1920s to document Maori traditional ways, remarked that the kea was no good for food, as it was too *maieke* (lean).[4] Other than the name, then, there is relatively little physical or cultural evidence to link the Maori with keas.

Ngai Tahu travelers from the South Island regularly crossed most of the major passes of the Southern Alps in search of *pounamu* (greenstone or nephrite jade), which they used to construct fishhooks, spear tips, and decorative and religious sculptures. So important was this stone that the whole of the South Island was known as "Te Wai Pounamu"—the water of greenstone. A chief source of the mineral was located to the west of what is today Arthur's Pass National Park, an area in which keas are now commonly observed. It is thus likely that the early Maori had contact with the birds. But their lack of influence on Maori culture suggests that they may have been rare, existing only in such small, isolated populations that the Maori had little use for them.[5]

Scientific findings support this notion. Although the remains of other New Zealand birds are common in Maori middens, kea bones and feathers are almost completely absent. Kaka bones are abundant in even the earliest of South Island middens, where they are often found with the bones of moas and other forest birds.[6] In contrast, the only midden site thought to contain kea remains is from Lee Island in Fiordland, on the southwest coast of the South Island. Among a large sample of kaka and kakapo feathers were two feathers identified as having come from keas. The midden in which they were found was radiocarbon dated from the sixteenth to the eighteenth century but also contained a small amount of twentieth-century material. If the feathers were correctly identified and are not of recent origin, they constitute the only direct, physical evidence of Maori contact with keas before European settlement.[7]

The South Island. The asterisk marks the location of the authors' study site, near Arthur's Pass Village. (Map by Bill Nelson)

SECOND WAVE: EUROPEAN SETTLEMENT

The first documented European encounter with the kea came fully seventy years after the discovery of its relative, the kaka. When Captain James Cook first visited New Zealand in 1770 in the *Endeavour,* the ship's log noted encounters with parrots on both the North and the South Island, almost certainly kakas. On Cook's return voyage in 1773, his men shot several South Island kakas at Dusky Sound in Fiordland, one of which was later painted by George Forster. Kakas were also seen on Cook's final visit to New Zealand in 1777; his ship's surgeon wrote of numerous "large brown parrots with white or greyish heads."[8] In 1781 John Latham described the "Southern Brown Parrot" based on the specimens from Cook's voyages, giving the habitat simply as "New Zealand." Based on Latham's description, J. F. Gmelin classified the species as *Psittacus* (later *Nestor*) *meridionalis* in 1788.[9]

When keas finally did come to the attention of European explorers, they were not recognized as a new species. A small group of French colonists arrived on the South Island just after British sovereignty had been declared in 1840. Their leader, Sainte Croix de Belligny, a naturalist from the Jardin des Plantes in Paris, collected over 130 species of birds, including keas, and sent the specimens back to France.[10] But keas receive no mention in the accounts of other contemporary scientific expeditions, such as the lengthy explorations of the ship *Acheron* in 1848 or the visits of either of England's most famous naturalists—Charles Darwin, who visited New Zealand in the *Beagle* in 1835, and T. H. Huxley, who visited in the *Rattlesnake* in 1850. Only in 1856, when Walter Mantell forwarded two specimens to the English ornithologist John Gould, were keas formally described. Gould characterized the species as *Nestor notabilis* in the *Proceedings of the Zoological Society* and included it in the supplement to his *Birds of Australia.*[11]

In the 1860s resident New Zealand naturalists considered keas to be rare. Sir Walter Buller, the founding father of New Zealand ornithology, first encountered them in 1860, on a ranch near the Rangitata Gorge; he described the bird as "essentially a mountain species, inhabiting the rugged slopes of the Southern Alps and descending to the plains only during severe winters when its customary haunts are covered with snow and its means of subsistence have consequently failed." Buller claimed that for many years the kea ranked among New Zealand's rarest species. Nonetheless, there clearly were restricted areas of abundance. In 1861 and 1862, during a topographical survey of the mountains of Otago, Sir James Hector found keas everywhere. They were apparently so plentiful and so tame that he had "no difficulty in knocking them over with a stone whenever he wanted to replenish his larder."[12]

Soon after the species was first recognized, however, kea populations seem to have exploded throughout their range. Buller now acknowledged that although keas were still very rare in the northern parts of the South Island, they had quickly become pests in the middle and southern districts.[13] Their expansion in numbers appears to have coincided with the arrival of European settlers in the high country.

Intensive settlement of the South Island high country began in the 1840s and 1850s. Runholders, as the settlers were called, could obtain inexpensive leases over areas large enough for up to twenty thousand sheep. Typically the tenants first burned off the native mixed shrubs and grasses from their large tracts in the fall and then returned with merino sheep from Australia in the spring, when the replacement grass was green. By 1870 such sheep runs were well established, producing and exporting fine wool.[14]

Sheep ranching on the South Island never offered the easy profits that were available elsewhere in the country, however.

The hardships of making these runs succeed were substantial, and few of the original runholders had any experience in sheep management. Even after burning, the tussock grass could support, at best, only one merino sheep per acre, and more commonly one sheep per four to five acres. In contrast, the better runs of the North Island could support up to five sheep per acre, as well as some cattle.[15]

To make matters worse, wool prices were fluctuating wildly. The American Civil War had created a high demand for imported wool when cotton supplies from the southern states were cut off. After the war, when cotton again became widely available, wool prices in New Zealand plummeted. By 1867 nine out of every ten runholders in the Marlborough District were virtually bankrupt, loaded down with mortgages on which they were paying up to 15 percent interest. Over the next seventeen years, as the sheep industry grew in the North Island, production declined in the South, and the average size of sheep flocks decreased by one-third.[16] To reduce labor costs, South Island flocks were allowed to wander at will through the run for a large part of the year. That practice, combined with the custom of disposing of sheep carcasses and offal in uncovered pits, opened a window of opportunity for the kea.

Keas frequented the high-country sheep runs; they seemed to live almost exclusively on carrion. A contemporary witness described keas feeding on sheep's heads thrown from the slaughter shed: "Perching itself on the sheep's head, or other offal, the bird proceeds to tear off the skin and flesh, devouring it piecemeal, after the manner of a Hawk, or at other times holding the object down with one foot, and with the other grasping the portion it was eating, after the ordinary fashion of Parrots."[17]

From Relict to Renegade / 33

Nineteenth-century illustration of a kea attacking a sheep in the snow. (From T. H. Potts 1976)

By 1867 sheep in the high country runs seemed to be afflicted by a new disease, manifested by the sudden appearance of a patch of raw flesh, about the size of a man's hand, on the loin area. Soon, however, an observant shepherd noticed a kea clinging to the back of a sheep and pecking at one of these sores.[18] Witnesses subsequently reported seeing keas loitering around sheep that were bleeding from fresh wounds. Some sheep did not have wounds but simply bare patches where tufts of wool had been picked out. The stomachs of keas shot on these sheep runs were often filled with wool and raw mutton.

Attracted by the sheep as well as by the accumulated carrion in high-country refuse dumps, keas were now daily visitors to places where they had only occasionally been sighted just a few years before. Their numbers had also increased; where once they had been seen in tens, now they could be seen in fifties.[19] Many sheep ranchers attributed the increase to the plentiful supply of food obtained from dead animals. By the late 1870s keas were widely considered to be abundant. They were common in the higher mountains of Southland, where they appeared to wander from place to place in small flocks of a dozen or more. It seemed the parrots would attack wherever sheep were being gathered or driven along. Then, after a few days, they would disappear and would not be seen again in the area for days or even weeks. On the lower ranges, under 600 m elevation, keas were only occasionally seen. By 1895 the numbers of kakas had declined, mostly because of the destruction of the lowland forests, but keas were no longer thought to be rare anywhere within their range.[20]

Their range, moreover, appeared to be extending in all directions. For the first few decades after their formal classification in 1856, keas were recorded from only the southern half of the South Island as far north as Arthur's Pass. By the turn of the century, however, keas were being recorded from the lower, forested mountains near the northern end of the South Island, as well as the eastern foothills of the Alps, adjacent to the Canterbury Plains.[21]

By the last decades of the nineteenth century, inexperience, poor soil, and low wool prices had taken their toll on the high-country sheep runs of the South Island. The additional burden of extremely cold winters and introduced pests forced many runholders out of business. Between 1873 and 1882 yearly exports of rabbit skins had increased from 33,000 to 9 million. The introduced rabbits caused widespread destruction to pastureland, which seriously affected the sheep-carrying capacity of the runs. At the Castle Rock ranch in Southland, for example, an area of seventy thousand acres had supported fifty thousand sheep before it was invaded by rabbits. After the rabbit invasion sheep stocks were progressively reduced until they stabilized at twenty thousand.[22]

One high-country Canterbury ranch of about thirty-eight thousand acres, the Clayton Run, reported that rabbits, wild pigs, and keas had been a great nuisance for many years. During the winter of 1885 runholders killed a hundred rabbits, sixty pigs, and three keas, and that spring there were apparently as many wild pigs as rabbits. In the severe winters of the 1880s the run lost from two to ten thousand sheep each winter, out of a total of about thirty thousand. The ranchers observed that keas gave them the most trouble during the most severe winters.[23]

In the face of heavy pressure from the runholders, the New Zealand government instituted a bounty system on keas, offering up to three shillings a head. In March of 1884 the sheep inspector at Queenstown reported that keas had attacked a flock on a ranch near Lake Wanaka and killed two hundred sheep in a single night. The beaks of 1,574 keas were later delivered to his office for payment of the reward.[24]

The easy availability of bounty pay, in an otherwise destitute economy, created a cottage industry in accumulating kea beaks. At first shepherds and musterers were content to carry guns while going about their work, taking the opportunity to shoot any keas they encountered. As keas became warier and harder to find, sheep ranchers began to promote direct hunting of the birds, offering as much as two shillings per kea head. In the end, exterminating keas became the province of professional bounty hunters, employed by sheep ranchers solely to seek out and kill keas and rabbits.

Bounty hunters received all their supplies from the ranchers, as well as a weekly wage and an extra fee for every kea head brought in. A one-handed .410-gauge shotgun is still known in New Zealand as a "kea gun," because it was often the hunters' weapon of choice. They used a wide range of methods to lure the birds: they displayed bright pieces of cloth, chained tame keas to rocks to "call in" other birds, and even taught themselves how to imitate the kea's call. Beginning at the turn of the century, hunters would also set out sheep carcasses poisoned with strychnine. Although discouraged on some ranches because of the risk of unintentionally killing sheepdogs, the practice continued for years.[25]

Allegations of sheep damage from keas varied enormously. An 1883 report claimed that keas had killed over fifteen thousand sheep in a single region. One Otago farmer insisted that keas had killed thirty thousand sheep on his run over a period of fifteen years, and he boasted of killing over three thousand keas. In 1908, seeking to assess the actual extent of kea damage to sheep, George Marriner wrote many of the runholders directly to document their views. Although some shepherds put the annual loss of stock from keas at 30 to 40 percent, he concluded that they probably exaggerated, and that 5 percent would be a more realistic estimate.[26]

By the end of the 1880s keas were closely associated with sheep in the minds of the New Zealand public. The 1888 edition of Walter Buller's acclaimed *History of the Birds of New Zealand* contains a print by the Dutch-born illustrator J. G. Keulemans that shows a kea perched on an overhang, while below it two other keas harass a sheep. A recent author has suggested that the distant birds and sheep in Keulemans's plate were Buller's attempt to "pander to the runholders' lethal prejudice against the species."[27]

Once considered a peaceful fruit and nectar feeder, the kea

Lithograph of the kea by J. G. Keulemans, from the 1888 edition of *History of the Birds of New Zealand* by the New Zealand naturalist Walter Lawry Buller. (From Turbott 1967; used by permission of Penguin Books [N.Z.] Ltd.)

seemed transformed, almost overnight, into a dedicated carnivore, rending the flesh from defenseless sheep throughout the high country. In 1871 the prolific New Zealand naturalist T. H. Potts published the first account of this transformation in the British journal *Nature:*

From the recent settlement of the country, it would be quite possible to date each step in the development of the destructiveness of the Kea, the gradual yet rapid change from the mild gentleness of a honey-eater, luxuriating amidst fragrant blossoms when the season was lapped in sunshine, or picking the berried fruits in the more sheltered gullies when winter has sternly crushed and hidden the vegetation of its summer haunts. Led, perhaps, to relish animal food from its partly insectivorous habits, its visits to the out-stations show something like the bold thievery of some of the Corvidae, whilst its attacks on sheep feeding on high ranges exhibit an amount of daring akin to the savage fierceness of a raptorial.[28]

The mechanism responsible for this conversion of habits was the subject of intense speculation. Most naturalists simply saw the kea's inherited predisposition for feeding on insect larvae as now generalized, in some fashion, to large, woolly mammals.[29] The difference in foods, they suggested, was more apparent than real. "Possibly," noted the New Zealand biologist W. B. Benham, "there is not a great amount of difference in taste between a good, fat, juicy weta or beetle grub and a piece of raw sheep."[30]

Probably the most popular conception was that keas had first been attracted by maggots on sheep carcasses: "Suppose that these birds formerly fed chiefly on berries and the large white grubs abounding in mossy vegetation on the hills; and that after the country was stocked they, first by feeding on maggots and insects on dead sheep, and afterwards on the dead animals, acquired not only a taste for meat, but also a discrimination of the choice parts."[31]

Other, more fanciful accounts were offered, including one that became known as the Vegetable Sheep Theory.[32] "Vegetable

Sheep" was the common name given to two genera of plants that grow on rocks above 1,200 m in the northern half of the South Island. These thick, low shrubs are covered with light-colored woolly leaves and bear a vague resemblance to a sheep in repose. The advocates of this theory suggested that keas were in the habit of tearing open these plants to get at the large white grubs that were said to live in them. Thus, when sheep first wandered into kea habitat, they were mistaken for plants. "The bird, with the intention of digging out the grubs, was supposed to tear open the animal's skin, and, finding meat and fat even more appetizing than the grubs, persisted in its efforts and so acquired the habit of sheep killing."[33]

Several observers suggested, however, that the kea's change in lifestyle probably resulted from a general adaptive ability, manifest in the bird's intelligence, curiosity, and mischievousness. That keas were unusually intelligent was well recognized. One nineteenth-century writer even credited them with the ability to communicate ideas among themselves.[34] He claimed to have observed several keas, after an apparent consultation, delegate one bird, twice in succession, to untie the knot in a string that fastened another of their number to a pick handle. For birds that could plan an escape from captivity, exploiting a novel food source seemed no great challenge.

Even in the 1860s, naturalists had remarked on the kea's unusual exploratory behavior. Buller relates that during a botanical survey, Sir Julius Haast once left "a large bundle of valuable alpine plants lying exposed on the summit of a lofty mountain crag. During his temporary absence a Kea came down and tumbled the collection of specimens over into the ravine below, and quite beyond recovery."[35] Keas were known to be extremely persistent and thorough in their destructiveness. One account cited a shepherd who, upon returning to his hut after an absence of a day or two, found a kea that had gained access by the chimney: "Blankets, bedding, and clothes were grievously rent and torn, pannikins and plates scattered about; and everything that could be broken was apparently broken very carefully, even the window framing having been attacked with great diligence."[36]

The initial rarity of keas, their rapid population increase associated with carrion, and their extraordinarily persistent and destructive exploratory behavior all suggested that sheep predation, far from being an aberration, was simply another manifestation of the pervasive curiosity of the species. Benham remarked in 1906, "There can be no doubt that the origin of the [flesh-eating] habit is traceable to the kea's natural curiosity: its bump of inquisitiveness is very highly developed, and it will investigate any unusual object—turning it over, pecking at it, and so forth." Even Walter Buller seemed convinced that "it is their insatiable curiosity that has probably led them to find out the taste of mutton."[37] This, in the last analysis, is the most tantalizing inference from the historical record: the possibility that for keas, foraging is a wholly open and opportunistic process, that the bird's diet is the result of essentially trying everything and then keeping what works.

IN THE WAKE OF SETTLEMENT

Since the first arrival of humans in New Zealand, much of the native forest has been burned or logged off. Maori hunters burned large areas to flush out and concentrate game. European colonists, driven by a desire for pastureland and timber, eliminated much of what was left. Temperate rainforest had once covered over three-quarters of the land surface of New Zealand; by the early 1900s it was reduced to less than a quarter.[38] Where trees were later replanted the new forests consisted of exotic species, such as Monterey pines, that few native birds could use.

The loss of the forests was not the only ecological catastrophe to be inflicted on New Zealand. Along with the waves of settlement came alien plants and animals, species that were completely new to the New Zealand ecosystem. Polynesian immigrants brought in dogs and *kiore,* or Polynesian rats. The real surge in introductions came during the European period, however. Whether for food or for sport, for the sake of familiarity or simply by accident, nineteenth-century European settlers introduced an additional 143 exotic animal species, about 34 of which remain established today, along with over 1,600 alien species of plants.[39] The frequency of introductions mirrored the intensity of settlement. Most of the species were introduced between 1850 and 1890, the most concentrated period of European immigration.

The introduced animals had as devastating an impact on the native fauna as did the destruction of the forests. Norway and black rats, stoats, dogs, and cats, finding many of the forest birds to be easy prey, decimated the bird populations. Along with the predators came a host of browsing mammals, including the Australian brush-tailed possum, the European rabbit, and a variety of wild deer. Largely introduced for sport hunting and the fur industry, these species caused serious damage to forests and alpine meadows and brought about extensive changes in the native vegetation.[40]

The kingdom of the moa was gone forever. The last of the

moas died in the first wave of human immigration to New Zealand, long before Captain Cook sighted the islands. The fossil record of all eleven species terminates at about 1600. The Maori hunted moas extensively, and as the human population increased, so did the pressure on the moa populations. Forest burning may have devastated remnant moa flocks that had already been weakened by overhunting. Once the forests were opened by fire, their structure may have been so altered that they no longer supported foods that were essential to native birds.[41] When the moas vanished, the large birds that depended on them for food, such as eagles, goshawks, and ravens, also disappeared. Small flightless birds, such as wrens and snipe-rails, were probably eliminated early on by *kiore*. Species such as kakapos declined dramatically after stoats became abundant in their habitat.[42]

The combined pressures of hunting, deforestation, and the introduction of predators and competitors resulted in a biological holocaust. Over the past thousand years, forty-three species of native birds have become extinct, including over a third of the original land birds (table 2). Of the native bird species that still exist, many are rare or confined to a few relict populations. Today New Zealand ranks with the Philippines as having the world's highest percentage of threatened birds (15 percent).[43]

Keas had existed as a relict species since the late Pleistocene, persisting at low population levels in marginal alpine habitats. In large part they survived the mass destruction wrought by human settlement. When the forests were burned and the moas eliminated, keas dropped carrion from their diets and shifted to other sources of food. As dietary generalists they were relatively resistant to the environmental changes that forced so many other birds, including their competitors, into extinction. When sheep

TABLE 2. New Zealand Birds Extinct in
the Last Thousand Years, Listed by Family

Common Name	Scientific Name	Date of Extinction
EMEIDAE (Moas)*		
	Anomalopteryx didiformis	
	Emeus crassus	
	Euryapteryx geranoides	
	E. curtus	
	Megalapteryx didinus	
	Pachyornis australis	
	P. elephantopus	
	P. mappini	
DIORNITHIDAE (Moas)*		
	Dinornis giganteus	
	D. novaezealandiae	
	D. struthoides	
PELECANIDAE (Pelicans)		
pelican, New Zealand	*Pelecanus novaezealandiae*	
ARDEIDAE (Herons, Egrets, and Bitterns)		
bittern, little	*Ixobrychus novaezelandiae*	1900 (Falla et al. 1978, 76)
ANATIDAE (Waterfowl)		
duck, Finch's	*Euryanas finschi*	
merganser, Auckland Is.	*Mergus australis*	1905 (Falla et al. 1978, 89)
swan, New Zealand	*Cygnus sumnerensis*	
	Biziura delautouri	
	Cnemiornis calcitrans	
	C. gracilis	
	Malacorhynchus scarletti	
	Pachyanas chathamica	
ACCIPITRIDAE (Raptors)		
eagle, Haast's	*Harpagornis moorei*	
eagle, sea	*Haliaeetus australis*	
goshawk, New Zealand	*Circus eylesi*	
PHASIANIDAE (Game Birds)		
quail, New Zealand	*Coturnix novaezelandiae*	1875 (Falla et al. 1978, 92)

(*continued*)

TABLE 2 *(continued)*

Common Name	Scientific Name	Date of Extinction
RALLIDAE (Rails, Gallinules, and Coots)		
adzebill	*Aptornis otidiformis*	
rail, Chatham Is.	*Gallirallus modestus* = *Rallus modestus*	1900 (Falla et al. 1978, 96)
rail, Dieffenbach's	*Rallus dieffenbachi*	before 1900 (Falla et al. 1978, 96)
rail, Hodgen's	*Capellirallus hodgeni* = *Gallinula hodgenorumsis* = *Gallinula hodegnorum*	
snipe-rail	*Capellirallus karamu*	
	Diaphorapteryx hawkinsi	
	Gallirallus minor	
	Nesophalaris chathamensis = *Fulca chathamensis*	
AEGOTHELIDAE (Owlet-nightjars)		
owlet-nightjar	*Aegotheles novaezealandiae*	
STRIGIDAE (Owls)		
owl, laughing	*Sceloglaux albifacies*	1914 (Heather and Robertson 1997, 366)
ACANTHISITTIDAE (New Zealand Wrens)		
bush wren	*Xenicus longipes*	1972 (Heather and Robertson 1997, 373)
bush wren, Stephen's Is.	*Traversia lyalli*	1894 (Heather and Robertson 1997, 371)
	Pachyplichas jagmi	est. 1,000 years B.P. (Millener 1988)
	P. yaldwyni	est. 1,000 years B.P. (Millener 1988)
SYLVIIDAE (Old-World Warblers)		
fernbird, Chatham Is.	*Bowdleria rufescens*	1900 (Heather and Robertson 1997, 386)
CALLAEIDAE (Wattlebirds)		
huia	*Heteralocha acutirostris*	1920 (Heather and Robertson 1997, 419)

(continued)

TABLE 2 *(continued)*

Common Name	Scientific Name	Date of Extinction
PARADISAEIDAE (Birds of Paradise)		
piopio	*Turnagra capensis*	1902 (Heather and Robertson 1997, 422)
CORVIDAE (Crows and Jays)		
raven, New Zealand	*Corvus moriorum* = *Paleocorax moriorum*	

NOTE: These forty-three species have all become extinct since humans first arrived on the islands. The dates indicate the last occurrence of a species, when known, followed by the reference. Extinct subspecies of species that are still extant have not been included.

REFERENCES: Falla et al. 1978; Ornithological Society of New Zealand 1970, 77–79; Millener 1988, 1990; Cooper et al. 1993; Cooper et al. 1992; King 1984, 217–20; Heather and Robertson 1997; R. N. Holdaway 1989a; Steadman 1995; J. M. Diamond and Veitch 1981; Bull and Whitaker 1975. Classification follows the convention of Heather and Robertson 1997.

*These two families comprise the eleven species of moa that are currently recognized (Millener 1990; Cooper et al. 1993; Cooper et al. 1992).

began to die in the snowfields, keas rediscovered a lucrative livelihood as scavengers. A rich resource had suddenly been injected into their habitat, and the other native species that might have shared in the spoils had long ago been driven to extinction. Their numbers increased dramatically, and the kea rode the roller coaster of fortune back to its apex. This ability to tolerate massive environmental change and make the most of new opportunities sets the kea apart from nearly every other island species.[44]

THREE

Hanging Out with the Gang

> From his roost on Avalanche Creek [the kea] looked straight down to the village of Arthur's Pass. Each morning at daybreak he would see and hear [other] keas feeding at the dump, and he would plunge down to join them.
>
> J. R. Jackson, "Life of the Kea"

Beginning in 1986 we conducted a field study of the kea in Arthur's Pass National Park on New Zealand's South Island (see map, p. 29). For the three following years we revisited the park, each time in the early summer of the Southern Hemisphere, from November through January. Our primary study site was the area surrounding a refuse dump near Arthur's Pass Village that keas have frequented for at least forty years.[1]

We chose this location in part because of an established tradition. Thirty years before, J. R. Jackson, a ranger with the New Zealand Park Service (now the Department of Conservation), had made this area the primary focus of the first major field study of the kea. During a decade of research Jackson banded more than six hundred keas, observed thirty-six active nests, and recorded voluminous notes. His series of five papers, published in the 1960s, laid the foundation for all future work on the field biology and behavior of the kea.[2]

Jackson spent long hours studying the birds at the dump site, tolerating the stench of garbage, the smoke when it was burned, and the vicious swarms of blackflies it attracted. His dedication became the stuff of legend. One anecdote concerns a park employee who had stopped at the site to dispose of his household garbage.[3] At the edge of the deep trench of accumulated garbage, he noticed a large burlap sack. Since his hands were full, he tried to kick the sack into the pit. One can imagine his surprise when out of the sack crawled Jackson, who had been hiding inside it to observe the keas.

Through his diligent efforts, Jackson established many important features of the wildlife ecology and population dynamics of the kea. Before we began our study, however, little was known about the birds' social behavior or their process of learning about the world. Jackson's field notes might have contributed valuable information on some of these issues. Unfortunately, they are no longer available. Jackson vanished on a camping trip in Westland National Park in 1989. Although searchers located his last campsite and scoured the surrounding forests, no evidence of his fate emerged, and his original field notes and documentation were later destroyed.

THE STUDY SITE

The 95,000 hectares of Arthur's Pass National Park straddle the main divide of the Southern Alps, the chain of rugged mountains that dominates the western third of the South Island. The topography ranges from mountain peaks, covered at the highest elevations with permanent snow and ice fields, to the floors of steep glacial valleys and the gravel flats of braided rivers.[4]

The abrupt rise of the Alps produces a rain shadow, dividing the region into contrasting zones of southern beech forest on the eastern side and a mixed rainforest of conifers and broad-leaved trees on the west. The park also supports a unique assemblage of smaller plant communities, which include subalpine shrub, tussock grassland in the alpine zone, alpine shrub and herb fields, alpine cushion bogs, and alpine scree.[5] Keas occupy a broad ecological niche, taking advantage of resources within all these habitats.

The park is known for its wide daily and seasonal variations in temperature and precipitation. On the average, rain falls about every other day, much of it in brief, high-intensity storms. The incidence is very erratic, however. When we arrived in 1988, there had been thirty-six continuous days of rain. After a respite of eight days of sun, the downpour resumed. The water soaks rapidly into the gravelly soil and leaches into rivers and streams. Following a heavy rain the Bealey River may rise as much as two meters in just a couple of hours, then fall back to normal levels within a day.[6]

Temperatures on the valley floors are subject to severe fluctuations. At Arthur's Pass Village in midsummer, the mean daily maximum temperature is only about 20°C, although very hot and dry periods do occur. Snow may fall above 1,500 m even in midsummer, and avalanches are a common hazard on the mountain slopes. In winter the average daily minimum is −2°C, and the valley floors experience about three or four heavy snowfalls per year. Even in the milder times of year the heavy rainfall makes thermal stress a significant problem. Every summer a few unprepared hikers get drenched in these mountains and develop hypothermia.[7]

The refuse trench at the study site is newly dug each year into the loose gravel of the valley floor and surrounded with spoil ridges about three meters high. The tracks of the transalpine railroad border the site on the east. Beyond the tracks flows the Bealey River, a broad, torrential stream choked with glacial till. On the south and west sides of the site mountain beech forest rises in an unbroken canopy to the ridge crest above. One of the five species of southern beech that clothe most of the high mountain slopes of New Zealand, mountain beech is probably the toughest variety, growing mainly in drier conditions in poor alpine soils. Beech forests are structurally simple: the canopy is dominated by beech trees, and few shrub species grow in the understory.[8]

The park is administered by the New Zealand Department of Conservation, whose staff in Arthur's Pass Village provided us with assistance of every kind. They gave us weather data, documented their sightings of birds we had banded, reassured villagers who wondered why we had moved into their dump, and shared daily information about keas and the other animals at the site. The dump, one of the few unregulated ones in the county, had become a sensitive management issue for the park administration. They were particularly interested in any information we might obtain concerning the impact of dump foraging on the kea population.

OBSERVING KEAS

Our study procedures at the site soon became routine. Coated with insect repellent to ward off the swarms of blackflies, we forged out from our van to capture, band, videotape, and photograph keas and record their vocalizations. We observed their behavior from inside the van, to minimize any disturbance to the birds. We generally parked within four or five meters of the newest refuse trench, so that birds both in the nearby beech trees and at the bottom of the trench would be visible.

For the first several days of each season we captured and banded keas at the site using a manually operated drop net. The trap was baited with butter, an irresistible delicacy for keas. The birds quickly developed ingenious methods for stealing the bait without getting caught. For example, one bird would simply grab the side of the trap and shake it until it dropped, then run over and grab the butter through the mesh. Other keas would wait until one of their comrades had sprung the trap. While we were running out to collect the captive, these lurking individuals would race over to try to extract the bait before we arrived. Some particularly nimble birds became experts at the "fast runthrough," racing under the net, grabbing the butter, and running out before the trap was sprung. After a day or so of trapping, the local keas became very wary and required progressively more sophisticated strategies, such as changing the appearance of the trap by switching the color of the net or moving it to a new location.

By the end of the study we had captured and marked fifty-two individuals. Each captive kea was banded with numbered Monel metal bands, as well as colored plastic bands to allow recognition of individuals at a distance. Our banding program was part of a larger effort organized by Kerry-Jayne Wilson of Lincoln University in Canterbury, under the auspices of the Department of Conservation. During the period of our study she coordinated all banding of keas in the central area of the South Island, thereby avoiding possible duplication of color combinations. From researchers working in nearby areas we received information about keas that appeared at our study site with unfamiliar band combinations. Likewise, we were notified when birds that we had banded were sighted at other locations.

After banding, we measured and weighed each kea and determined its age and sex. Unlike most parrots, keas show a consistent change in color patterns as they age.[9] Fledglings, in the summer of their emergence from the nest, have a distinctive, conspicuous plumage. Their eyes are ringed with bright yellow-orange, and a similar color infuses the lower bill and the cere, the turgid region around the nostrils. In their first two months fledglings also have a light yellow cast to their crown feathers, a feature that can be discriminated at a distance even under low light conditions.

Except for the crown feathers, the fledgling color pattern fades only gradually over time, making it possible to identify two other classes of young birds. Juveniles, which are birds in their second summer, have a pale yellow eye ring, cere, and lower mandible. Because they have not yet molted, their feathers are often very dull and worn. Later, as subadults, their yellow eye ring is no longer complete. They may still have traces of yellow in their ceres, but their bills have become dark. The loss of cere and bill coloration seems to be associated with the first molt, which suggests that subadults are usually birds in their third or

Fledgling keas can be identified by the coloration of the eye ring, the cere (the area around the nares above the bill), the lower bill or mandible, and the light yellow cap. In the adult, shown opposite, these areas are all dark. (Photographs by Judy Diamond)

fourth summer. In adults the eye ring, cere, crown, and bill are all dark brown.

When we began our study, the existing literature provided only inconsistent and contradictory criteria for determining the sex of keas. Our first investigation of the species, therefore, was an analysis of sexual dimorphism using measurements from 224 keas—museum specimens, wild caught birds, and captive keas from the San Diego Zoo. We found that female keas can be

distinguished from males on the basis of bill length and overall size. Males are about 20 percent heavier than females, and their upper bills (culmens) are 10–15 percent longer.[10] Once attuned to this difference, we could accurately identify the birds' sex by eye in the field.

Over the course of our study we logged a total of 450 hours of field observation of keas. We generated a detailed behavioral database consisting of one hour of accumulated observation time for each of sixty-five identifiable individuals, including several that had been banded by other researchers. We recorded behav-

The authors studied kea skins and skeletons from museums around the world to determine the distinguishing characteristics of males and females. Here the length of the upper bill, or culmen, is being measured. (Photograph by Judy Diamond)

ioral events as keystrokes on a laptop computer, and the program kept track of the time intervals between successive events with an effective accuracy of tenths of a second. During each sample period all actions of the sixty-five focal individuals were recorded, along with the identity and behavior of any birds that interacted with them. We recorded forty-five distinct behaviors, based on an ethogram that we developed in the field during an initial pilot study in New Zealand and then refined through observations of captive keas at the San Diego Zoo. This database constitutes the largest accumulation of quantitative information on kea behavior in the field.[11]

Hanging Out with the Gang / 55

Male and female keas *(above and below, respectively)* can be distinguished in the field by the length of their upper bills, as well as by their overall size. (Illustration by Mark Marcuson)

Each season, after completing banding, we conducted a series of hourly counts at the study site, recording the total number of birds present, their age and sex classification, their location and mode of activity, and the identities of any banded individuals. Over three years we took a total of 459 such censuses. We found that an average of sixty-six different birds used the site in any one year, about half of them adults and the rest subadults, juveniles, and fledglings. The numbers of adults and subadults were quite stable across years, varying by only one or two individuals. Juvenile numbers, however, varied dramatically from one year to the next. Each year about seven new fledglings were added to the resident population.[12]

Keas gather in the beech trees at the study site several hours before sunrise. As they assemble, the birds begin to vocalize, producing variants of what we have called a "bleat-trill," and to move around in the trees. This ritual, played out for the past forty years in generations of keas, initiates a kind of kea convention at a place that provides a reliable source of rich food. In the course of feeding, or attempting to find food, the birds play out the dynamics of their social structure. While observing aspects of their foraging behavior, we also recorded their dominance hierarchies, their aggressive and affiliative behavior, and their social and object play. Most consistently striking in these observations were the distinctive roles played by the different age and sex classes in the kea community.

FORAGING

Keas seldom begin to forage on the ground until there is enough light to discriminate objects fairly well, usually at about 5:30 A.M. but as late as 6:00 or 7:00 when it is raining or overcast. Once the foraging session has begun, the activity level at the site is sustained through the early morning hours, with birds continuously arriving and departing. During peak periods there are often as many as twenty to thirty keas at the site.

The first bird to begin foraging is almost always a single adult, usually a male. This pioneering individual flies from the trees to the edge of the trench and then walks over to the refuse heap. Keas almost never fly directly to the dump. Once the first adult has begun active foraging, other birds follow suit. Juveniles, in particular, perch in the trees and call excitedly for long periods, waiting for an adult to make the first move. When no adults are present, one or two juveniles may come down and forage, but no others will join them. One adult on the site, however, breaks the ice and draws a squad of younger birds.

Adult male keas are capable scavengers, particularly adept at discovering and exploiting new food resources. They forage vigorously, with strong and coordinated movements, scanning the area with a practiced eye and focusing their attention on the most rewarding possibilities. For this enterprise they are equipped with a set of all-purpose foraging tools, the most essential of which is the bill. Although the kea's strongly curved upper bill looks superficially like that of a hawk or a kite, it is not sharply pointed but rather narrow and blunt, more like a screwdriver than an awl. Like a Swiss army knife, it can be used for an almost unlimited variety of operations: digging, pulling, scraping, prying, probing, and tearing. Keas also use their tongues in foraging to manipulate small objects or to abrade pieces of food against the hard palate, and like other parrots they use their feet to hold food or to brace items being chewed or scraped.[13]

Adult males engage in a broad array of foraging behaviors.[14] The most frequent activities we observed at the site include searching, which consists of wandering back and forth over the area, periodically touching and manipulating individual objects; and eating, that is, consuming concentrated foods that do not require additional processing. Because rich food sources are often not apparent on the surface, another common activity of adult males is excavating new resources by pulling off intervening layers of rubbish and throwing them aside. Occasionally, food discovery also requires demolishing a box or other container by repeatedly pulling, prying, and tearing to get it open. Keas often steal portable food items from one another and carry them away to eat under more secure conditions. In leaner times, adult males scrape fragments of meat from bones or residual flesh from soft fruits and glean small morsels of food that may be scattered broadly across the area.

When foraging, adult males throw superfluous objects left and right, often to a distance of several meters. Sometimes they even deliberately throw objects at juveniles who harass them. We saw one adult who had been plagued by two juveniles drive them off by repeatedly throwing rocks and garbage at them. He was surprisingly accurate, hitting one juvenile three times in succession over a distance of two meters.

Adult females spend less time eating at the dump than males of any age or even younger females. When they do forage at the site, they mainly glean small crumbs of food, scrape meat from bones, or steal food from other animals. Adult females may not be as dependent on food at the study site because, having recently fledged young, they are still being provisioned by their mates. They may also be more vegetarian in their preferences; we found that females spent more time than males feeding on the buds and leaves of the beech trees around the site. Or perhaps they simply do most of their foraging elsewhere. Throughout the three consecutive years that we conducted population censuses, only about 11 percent of the keas observed at the site were females.[15]

WHO'S IN CHARGE?

In keas, as in many other animals, the social organization of a group becomes manifest in aggression. The outcome of aggressive acts can indicate the relative status of individuals in the society and in some cases provide information about their priority in access to resources. These relationships of dominating and deferring generate an overall ranking system, sometimes referred to as a social hierarchy.[16]

Aggression among keas at the study site is frequent and intense, particularly among the young birds, since they lack an established position in the social hierarchy.[17] They strike with their wings and feet, attempting to knock their opponent off-balance. They bite any exposed area, feinting toward the face and eyes and pulling and twisting feathers and skin when they can reach them. The relationship between fighting and food resources seems only tenuous, however. Keas do squabble over high-quality foods, but much of the most vicious behavior appears wholly unmotivated by immediate gain. Adult females attack female fledglings and juveniles mercilessly, apparently irrespective of where they are or what they are doing. Adult and subadult males will break off foraging and launch an attack on a fledgling or juvenile that is simply playing on the other side of the site.

Full-scale aggression is less common among adult males, presumably because their relative priority has already been determined. At the approach of a more dominant bird, a subordinate kea usually defers by moving on and foraging at another spot. Adult males often displace each other with no displays of aggression. If an individual locates a highly desirable food, he may carry it away to where he can eat without disturbance. If the item is not portable, he can usually defend the find with no more than threats, such as gaping or feinting with the bill or striking at an opponent with one wing. Occasionally, two males of similar rank will face off over a choice morsel and engage in what we call a "stare-down." They stand, head to head, each with the tip of his bill a couple of centimeters from his opponent's eye, and then wait motionless, often for several seconds, until one participant loses his nerve and moves away.

Within kea society, males maintain a strongly enforced dominance structure. The highest-ranking birds are generally adult and juvenile males. Although one might expect subadult males to rank higher than juveniles, they are almost always displaced by adults, whereas juveniles are not. The dominance structure is not a linear pecking order, however, and relationships are seldom transitive. That is, if bird A dominates bird B, and bird B dominates bird C, that does not guarantee that bird A will also dominate bird C. For example, almost any adult can displace any subadult, and subadults frequently displace juveniles and fledglings. But some of these same juveniles will also displace some adults.[18]

Distinctive postures of the head feathers in keas appear to be linked to the motivation of the bird and its relative social status. These feather postures are, in effect, facial expressions. In the presence of high-ranking keas, for example, lower-ranking adults often fluff their head feathers while foraging. From the front their heads look soft and round, so we termed this the "owl face." It may indicate a mild defensiveness or apprehension. An actively aggressive kea fully ruffles its crown and nape feathers into a "hawk face," so-called because the ruffled crown creates a ridge above the eyes resembling that of a bird of prey. Birds

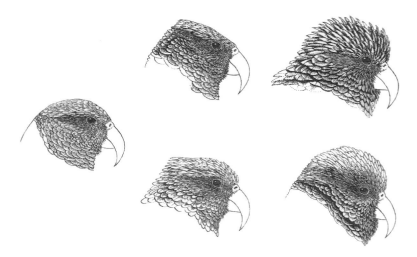

Keas show a range of characteristic facial expressions that communicate dominance status and aggressive motivation. *Clockwise from far left:* normal; "jay face," shown by dominant males in possession of a resource; "hawk face," an aggressive display; "owl face," an expression of defensiveness in subordinate individuals; and "nape erect," a submissive display. (Illustration by Mark Marcuson)

showing hawk faces are often attacked by others or initiate attacks themselves.

Confident, dominant birds usually show no particular change in feather orientation, but sometimes they sleek the anterior crown feathers slightly while raising the posterior ones. In profile the bird then looks as if it had a crest, like a blue jay or cardinal. We have therefore termed this expression the "jay face." Dominant males use this display while in possession of a resource. We have never seen it shown by more than one bird in a foraging group at a time. A submissive kea that is about to be attacked by another individual sometimes erects its nape feathers while sleeking the entire crown. This expression, which we termed "nape erect," is also part of the display produced by fledglings to request food or preening. Submissive birds showing this facial expression will often also sleek their body feathers and crouch.

Such facial expressions are strikingly analogous to those used by ravens competing over a carcass. Aggressive ravens show erect earlike tufts and often fluff out their throat feathers as well. Subordinate ravens fluff out the feathers at the top and sides of the head. With both keas and ravens, birds that have access to preferred feeding sites show dominant feather positions, while those watching and waiting for a feeding opportunity show subordinate ones.[19]

DANGER FROM ABOVE

A variety of other species share the kea's habitat at Arthur's Pass. Rock wrens, pipits, silvereyes, riflemen, and New Zealand robins can all be found in the alpine scrub, tussock, and beech forest surrounding the study site. In addition, house mice, feral cats, stoats, house sparrows, chaffinches, blackbirds, and black-backed gulls frequent the foraging site.[20] Keas are not particularly intimidated by any of them. Foraging keas generally ignore other species, except to engage in playful harassment. Juveniles occasionally chase stoats and cats, and they tease them by boldly pulling on their tails. Adult females, however, will launch repeated attacks on blackbirds, chaffinches, and even black-backed gulls. Although the gulls are nearly twice their size, female keas regularly drive them out of the area.

Keas remain vigilant when foraging in the open and periodically break off their feeding to survey their surroundings. On occasion a foraging kea will take alarm and fly to the trees with a warning cry, whereupon most of the flock will immediately rise and follow suit. They will remain in the trees, vocalizing, for several minutes until gradually they return to the ground.

During episodes of vigilance, keas continually scan the sky, which suggests that they are concerned with the possibility of attack from above. The only living aerial predator that would pose a threat is the New Zealand falcon, an elegant hawk roughly the size of a peregrine. Though uncommon, they are by no means rare in the high country and in forested mountain valleys.[21] These notoriously fierce and fearless predators are reported to have killed birds as large as guinea fowl, ducks, and pukekos, as well as many native forest species, such as tuis, pigeons, kakas, and parakeets.

In one incident we observed near the study site, a juvenile falcon perched about midway up in a large beech tree suddenly took off and began to attack several juvenile keas that were roosting near the top of the tree. They produced an unusual series of alarm calls, and an adult kea quickly emerged from the nearby forest and attacked the hawk. The falcon then turned on the adult, pursuing it through complex loops and spirals. Another adult kea joined the fray, and the two flew rapidly out of view, with the falcon in hot pursuit.

Other observers have also recorded falcon attacks on keas.[22] One author recounted an attack by a pair of falcons on four keas in an open area of tussock grass above tree line. Following a midair strike by one of the hawks, one kea staggered back, flapping and squawking wildly. Then both falcons attacked, flying about 15 m above the keas and swooping on them until the keas hid in a pile of rocks. After about ten minutes the falcons moved off.

Although falcons clearly will strike at keas, no observer has ever claimed to see successful predation. Keas may be vulnerable only when alone and exposed on open ground. The caution they display in descending to the ground to forage and their wariness with respect to potential aerial attackers suggest that the birds take pains to avoid being placed in this situation. An alert kea that is already in flight may be proof against falcon attacks, and keas do not hesitate to mob a falcon that is flying after other flock members.[23]

HANGING OUT

When the ground temperature begins to rise, usually around midmorning, keas cease foraging. By late morning most have left the site and do not return until later in the day. A few juveniles and subadults commonly remain in the trees near the site throughout the day. Keas appear to roost quietly and even to sleep through most of the early afternoon, usually within three to five kilometers of their feeding site.[24]

The actual hours of foraging depend strongly on the weather. During a series of hot, dry days, keas often leave the site just after sunrise and may not return until after sundown. But in cool, wet weather they forage until late into the morning and then return in midafternoon. Keas appear to have a relatively low thermal optimum, as might be expected of a species that has often been observed playing at high elevations in the winter snow. We have seen keas that were foraging at air temperatures of around

Hanging Out with the Gang / 65

A subadult male kea allopreens a juvenile male. (Photograph by Judy Diamond)

27°C flutter their throat pouches in an effort to dissipate excess heat. In essence, they were panting. Indeed, experiments show that keas begin to lose control of their body temperature when their surroundings are above 26 to 28°C.[25] Their pattern of feeding at dawn and dusk may be an adaptation to avoid the valley floors during the heat of the day.

Keas often fly to the surrounding beech trees to preen and socialize for a time before leaving the site. One conspicuous activity in this social interlude is allopreening, in which one bird grooms another by nibbling at its plumage. The head is the usual focus, with particular attention being paid to the feathers on the crown, nape, and throat and to the area around the eyes. The recipient may move its head to invite additional preening or crouch and ruffle its nape feathers in a submissive, soliciting pose.[26] Sometimes keas allopreen in groups, with up to three birds of different ages preening one another simultaneously.

More than simply idle behavior, allopreening is a social panacea, lubricating the rough spots in interactions and cementing attachments and allegiances. In the latter respect it resembles social grooming in apes or a handshake or hug between humans.[27] Keas allopreen in various combinations and contexts. Adult or subadult males allopreen younger birds for up to ten minutes at a time. During sexual interactions the male and female allopreen each other before the male feeds and mounts the female. Brief episodes of allopreening are scattered through the quieter interludes in extended play sessions, perhaps as a means of sustaining and reinitiating the interaction. Finally, allopreening may also take place more or less in passing, in very brief encounters during otherwise independent foraging.

Adult males apparently never allopreen other adult males, nor do females allopreen other females of any age. Keas in nearly all other combinations of age class and sex may allopreen, however. Furthermore, allopreening partners are quite consistent. The same pairs of individuals can often be seen socializing together in the same fashion for a month or more.

Juveniles frequently solicit allopreening; they can be most tenacious at sidling up to adults and presenting their crowns to be preened. Sometimes younger keas actively compete for the attentions of an older individual. One adult male was very attached to a particular subadult. The two birds roosted together, and the adult preened the subadult regularly for long periods.

An adult male kea regurgitates food to a fledgling. (Illustration by Mark Marcuson)

This adult also had two fledglings from the current season's nest, and the young birds constantly lurked nearby in postures of solicitation. Their presence was not appreciated by the subadult, who would often turn from his preening ecstasy to bite or kick a soliciting fledgling.

Kea fledglings are poor foragers. Like most other birds, they are fed by their parents, particularly their fathers, who regurgitate food to them until they are well on their way to independence.[28] To solicit either food or preening from their parents, fledglings adopt a characteristic pose, called "crouching," in which they lower both the head and the body toward the ground. They arch their necks to draw the bill down into the breast feathers. The nape feathers are erected and the throat feathers fluffed, creating a dark ring around the sleeked yellow crown that is quite striking when seen from the front. The body feathers are also partially fluffed. The wings are held close to the body and are not fanned open, and the tail feathers are partially spread. Oriented toward its parent, the fledgling stands motionless, sustaining the display posture. If the adult moves away the fledgling generally follows, staying close to the adult while still keeping its head and neck down. Fledglings seldom vocalize during this display.

Many components of this fledgling display are retained in the crouch posture of older birds, but over time the posture takes on new meanings. Fledglings and juveniles crouch to solicit food or preening. Subadult males and subadult and adult females crouch to indicate submissiveness in response to the approach of an adult male. Sexually mature females also crouch to solicit preening, feeding, and copulation from males. The semantics of the crouch display are thus quite complex. The behavior carries overtones of submissiveness, dependency, and sexual solicitation all at once.[29]

One incident we observed near the study site illustrates the consequences of this ambiguity. An adult male was perched near the top of a beech with two fledglings, one male and one female. All three birds were banded and known to us from earlier observations. We had been watching the fledglings solicit food and preening from this male for several days, which indicated that he was their father. The female fledgling approached the adult male in a solicitation pose. He preened her, then abruptly grabbed at her nape and mounted her. Instead of copulating, however, he climbed down again and rubbed his cheek back and forth along a branch, after which he fed the fledgling repeatedly. A short time later the male and the two fledglings flew off together down the valley.

This curious incident suggests that ambiguity in the significance of the crouch display may sometimes lead to motivational confusion. Regurgitant feeding is, after all, a sexual act for parrots, as well as a means of provisioning offspring. Male keas feed their mates as part of courtship and maintenance of the pair bond.[30] Perhaps the act of regurgitant feeding itself may engender conflicting motivations in a male parent, particularly in the case of female offspring that are fledged and approaching an age for pairing.

PLAYTIME

At the study site keas play at all times of the day and evening. Many kea displays are fairly subtle, but their play behavior is so intense and distinctive that it is impossible to overlook. Episodes of play typically persist for only a few minutes, but we occasionally saw bouts that lasted up to an hour. Keas play with many different kinds of objects and engage in elaborate social play, either in pairs or in groups of five or more. The prevalence and intensity of their play is unique among birds, as can be recognized from the variety of studies that have been conducted on kea play in captivity. And although various bird species, particularly parrots and ravens, are known to play, there are few detailed descriptions of the dynamics of avian play in the field.[31] This book provides the first detailed account of play among keas in the wild.

For young keas play is often a rough-and-tumble affair. They stand face-to-face and jump repeatedly, flapping their wings wildly. They push each other with their feet and wrestle with their bills, twisting their heads from side to side. One bird will roll on its back and the other will immediately leap on top, the two birds kicking and locking bills. Sometimes one kea will grasp the partner by the throat and drag him or her across the ground. Keas also chase each other, both on foot and in flight;

play bouts that begin on the ground will often continue unabated in the trees.

Arboreal play is spectacular, with the participants jumping on each other and hanging upside-down, often by one foot, while biting the playmate. A hanging bird appears to invite being jumped on and having its feet bitten, which forces it to let go and drop to a lower layer in the tree, where the game begins anew. Throughout this exhausting activity the play partners vocalize continuously, producing a remarkably diverse array of squeals, shrieks, and gurgles that are otherwise rarely heard.

Several characteristic play behaviors are sometimes seen in other contexts but appear to carry special meanings for play interactions.[32] For example, two playing keas often make repeated short, vertical jumps while furiously flapping their wings. Observers have likened this behavior to dancing, because the birds face each other and occasionally appear to jump in unison. Outside the play context, jumping and flapping may occur in the course of a fight as a component of wing-hitting, a strictly aggressive behavior similar in intensity to a serious bite. In this case, however, the action is seldom immediately repeated, and it is generally performed by only one bird at a time. When it occurs in play, the action is extraordinarily facilitative, appearing to reinforce and sustain the play activity.

Another typical play behavior is rolling over on the back while waving the feet in the air. Keas do this repeatedly during long play sessions, usually squealing at the same time. The other bird often runs over and leaps on the stomach of the displaying individual. Like jumping and flapping, rolling over is reminiscent of more serious kea behavior. In the midst of a full-scale battle, a kea will sometimes roll over on its back to fend off an attacker

A pair of fledgling male keas jump and flap during an episode of tussle play. (Illustration by Mark Marcuson)

by kicking.[33] In play, however, keas usually roll over when the play partner is some distance away, encouraging the partner, it seems, not merely to approach but to take a flying leap at the bird's exposed belly. We thus believe that rolling over may serve as an invitation, a means of inducing a partner to begin or to continue a bout of social play.

Another kea behavior that appears to carry a special significance in play is tossing a stone, stick, or other small object vertically in the air. It recalls excavation movements, where objects are thrown laterally to expose concealed food resources. Here, however, the object goes straight up and, as if to accelerate its ascent, the bird will often make a short hop to accompany the toss. Females may perform the action repeatedly over the course of a play interaction and continue to toss objects even after their part-

During tussle play between a pair of juvenile male keas, one bird rolls on his back with his feet in the air, and the other jumps on his stomach. (Illustration by Mark Marcuson)

ner, invariably a male, has moved away. We suspect that tossing serves as a female's invitation to courtship play, analogous to the manipulation of objects by crows and magpies during their precopulatory displays.[34]

Play signals like jumping and flapping, rolling, and tossing modify the significance of other behavior patterns and reassure the partner of the essential playfulness of the interaction. Play in keas thus seems to be fully analogous to play in mammals, particularly primates and canids, that also show play signals.[35] A play face in primates or a play bow in canids appears to change the meaning of subsequent actions, letting the participants know that this is all in fun. Similarly, the play signals of keas may allow more characteristically aggressive behavior patterns, such as kicking and biting, to be considered as playful rather than threatening and to be readily understood by both partners.

TUSSLE PLAY AND TOSS PLAY

Keas display two distinctive types of play. The first, tussle play, is essentially rough-and-tumble play. It comprises five primary behavior patterns: jumping and flapping, rolling over, biting, locking bills, and pushing with the feet. The last two are parts of a wrestling maneuver, in which one participant grasps the other's bill and twists it back and forth, often while pushing him or her in the stomach with one foot. Other, more characteristically aggressive, actions are often incorporated in this type of play, including gaping and feinting with the bill, biting the feet, and occasionally even taking the partner by the neck and dragging him or her around.

A striking feature of tussle play is that while any single bird's actions within the play sequence have no apparent pattern, the relationship between the partners looks almost choreographed. Each individual shows a strong tendency to echo what the other has just done. The amount of such facilitation within tussle play sequences is unusual when compared to other types of social interaction in keas. In some sequences almost two-thirds of the successive behavioral exchanges include components that are facilitated.[36]

The distribution of tussle play by age and sex follows complex rules. Tussling occurs primarily between fledglings of either sex, between fledgling and juvenile males, and between subadult or adult females interacting with fledgling or juvenile males. Juvenile females rarely play with other individuals of either sex. Tussle play is even less frequent among subadult males, although they occasionally interact in this fashion with juvenile males. Adult males rarely engage in tussle play, and when they do, it is mainly with mature females.[37]

The second type of play, toss play, appears to function in courtship or maintenance of a pair bond. Its chief actions are tossing, jumping and flapping, locking bills, and pushing with the feet. Somewhat less intense than tussle play, toss play seldom includes rolling over or biting. It is performed most frequently by adult and subadult females to solicit interaction with mature males. Although the males occasionally join in, the tossing action itself is mainly performed by females.[38]

PLAY CONFLICTS

Like other aspects of kea social life, play is occasionally enmeshed in motivational conflicts. A forty-five-minute episode we observed among three male keas—an adult, a juvenile, and a fledgling—illustrates these discordant interactions. The adult kea was feeding the fledgling, when the juvenile approached them and drew the fledgling away into a long play session that involved biting, interlocking bills, mutual jumping and flapping, and rolling over. The adult repeatedly approached the playing youngsters, trying to regurgitate to its offspring. Sometimes the juvenile would stop playing and chase after the approaching adult, who would dodge him and run off. At other times the adult managed briefly to feed the fledgling, although the juvenile continued to interfere. In one case, the fledgling was lying on his back and the juvenile stood on top of him, gripping him around the throat, while the adult shoved food into the fledgling's mouth. The adult never showed aggression to either the fledgling or the juvenile, but he always evaded the juvenile's approaches. Twice he tried to drag the fledgling away from the juvenile by pulling on the former's neck, but the juvenile simply followed the pair and continued his play attempts.

Two aspects of this odd interaction are noteworthy. First, the adult actively sought out the fledgling to feed him. The fledgling made no obvious effort either to solicit or to avoid being fed. Second, the juvenile repeatedly used play to interfere with the feeding, besides trying to approach the adult directly and to pry open the fledgling's bill. The juvenile's behavior suggests that he was attempting to obtain the food for himself.

Such conflicts between the motivations of adults and offspring may be common, much as when human parents try to get their children to stop playing and come to dinner. On another occasion we watched two juvenile males engaged in a ten-minute play sequence involving bill-wrestling. An adult female followed them around like a referee, at times biting the feathers of one or the other of the participants, but always staying within a few centimeters of the tussling pair. Perhaps she was attempting to break up the interaction or perhaps she just wanted to join in.

We have also seen several instances in which the mutual agreement for play may have been lacking. A typical case involved two juvenile males, one of whom repeatedly attempted to engage the other in tussle play. In response, however, the other juvenile would only crouch and solicit allopreening. The resulting behavioral sequence was quite confusing. At first the initiating juvenile would quietly nibble at the nape feathers of the soliciting bird. Then abruptly he would hop, flap his wings, and jump over his passive partner. Finally he jumped on the back of the solicitor, who then rolled over on his back. When the playful juvenile then jumped onto his stomach, the soliciting bird turned back over and, resuming a crouched position, nuzzled up to his partner. The initiating bird preened him for several seconds, hopped again, and then gave him a fierce bite. At this point the second juvenile ran away.

In addition to communicating readiness to play, play signals may inform others of the strength of an individual's interest in play, allowing play partners to negotiate the subsequent course of the interaction.[39] It seems likely, from our observations of kea play, that these negotiations may often be indeterminate. Certainly the pair of juveniles in our last example did not seem able to come to a consensus. The case of the juvenile that used play to interfere with an adult's attempts to feed a younger bird goes further: it suggests that play may, on occasion, participate in a confusing jumble of motivations and may even be used for ulterior, manipulative purposes. In this regard play mirrors the depth and complexity of other aspects of the kea's social life.

TOYS

Play in keas is less a set of ritualized behaviors than an attitude toward the world at large. Keas not only play with each other, but they are also perfectly willing to play with anything that offers a sufficient set of interesting properties. Social play for keas is thus only one aspect of a broader phenomenon that often incorporates the use of objects, much as for puppies or children play includes the use of toys.

The properties that make an object attractive to keas for play can be quite uncomplicated. We once observed a juvenile male engage in about twenty minutes of play with an anchored stick, treating it as if it were an opponent. He chewed on it, jumped on it, hit it with his wings, swung from it, rolled under it, and struck it with his feet, all the while vocalizing fiercely. When played with in a social context, almost any item of sufficient size will do, since the interaction with the partner provides the attractive element. A pair of fledglings will often manipulate a single object together—such as a stick, a bone, or a piece of cloth—pulling at it from both ends or repeatedly stealing it from each other in a variation on the classic games of tug-of-war and keep-away played by dogs and children.[40] Such object-oriented games occasionally develop into active tussle play, and the forgotten toy is left behind.

Although keas play with a great range of objects—from pieces of twig, bone, and flowers to larger items such as sticks and rocks—the selection of objects in solitary play is usually far from random. In general, play value is a function of what have been termed the "affordances" of the object, the intrinsic and contextual properties that determine the number of ways it can be used. The more affordances an object has, the longer a kea will play with it. Thus, juvenile keas in captivity find that a plastic cube with a slot on one side, which can be probed with the bill or peered into, is inherently more attractive than the same-sized cube without a slot. In the wild many of the most important affordances of an object are aspects of its context. Keas spend hours rolling rocks down inclines and throwing them into water, and clearly the prior position of the rock has a major influence on its play value.[41]

Keas are slow to tire of play objects, returning to the same ones over and over. The motivation to play is exceedingly powerful. Michaela Ritzmeier found that captive keas gave priority to investigating and manipulating attractive inedible objects, even when they were hungry and food was readily available in another part of the cage.[42] When a kea finally loses interest in an object, it can still be induced to return to it if another kea starts playing with it. Like many other aspects of kea social life, object play is highly facilitative. One kea playing with an object will often attract several more, until eventually a group of young birds are all excitedly tugging on the same item. We once watched a party of five or six juveniles play multidirectional tug-of-war with an old roll of gauze for the better part of an hour.

Object play in keas merges imperceptibly with other categories of object manipulation in which the element of play is less obvious. All parrots spend an enormous amount of time engaged in repetitious fiddling with small items, rolling them with their bills and tongues, much like a human chewing on a toothpick. Unlike most other parrots, however, young keas display a seemingly endless appetite for destroying large objects. Fledglings and juveniles spend many of their waking hours in demolition, ripping up old carpeting, discarded furniture, or even parked automobiles. Such manipulations are manifestations of the same general impulse that motivates object play.

Hanging Out with the Gang / 79

Three juvenile keas play tug-of-war with a piece of surgical gauze. Such games can often involve as many as five or six individuals. (Illustration by Mark Marcuson)

The similarities of demolition to object play are unmistakable. To begin with, demolition has only tenuous connections to foraging. Keas continue to tear apart objects even at the end of a foraging session, when they are presumably well fed. Their interest in objects for demolition is related more to the object's affordances than to its potential for providing food resources. Objects that are soft and yielding—for example, cloth and foam rubber—support a broader range of manipulations than harder items and are accorded greater interest. In addition, keas are extraordinarily repetitive and persistent in their destructive activities, just as they are in playing with any object. Finally, like other types of object play, demolition is strongly socially facilitated. Once one kea begins to pull and pry at an object, the activity invariably attracts others, eventually escalating into a concentrated play attack on the object by a band of juveniles and fledglings.

A KEA'S DAY ENDS

Late in the afternoon, beginning around 4:00, the number of keas at the study site sharply increases, as they resume foraging. They remain active in the area until it is quite dark, at which time the adults and their attendant fledglings leave. Compared to juveniles and subadults, adults spend relatively little of their time at the study site, and they generally do not linger when they are not foraging.[43]

The birds signal their departure with a rising chorus of "bleat-trill" vocalizations, which synchronizes the exodus of coherent groups. Abruptly several birds take wing, giving their trademark *kee-ah* call, and fly off down the valley. The flight call often triggers the departure of other individuals, even if they had not previously shown a readiness to leave. We once watched an adult leave the study site with a flight call, and at once six fledglings who had been industriously demolishing a chair dropped their game and took off after him. Sometimes the members of a pair vocalize antiphonally, exchanging low-level *meow* calls that synchronize their departure from the site. One member of the pair invariably takes off first, with a *kee-ah* flight call, and the other immediately flies after.

After dark the study site takes on a different atmosphere, becoming less like a family picnic and more like a street gathering. Most of these late-night visitors are juveniles and subadults that have stayed on after their early evening foraging session. In large part these young birds are not foraging but engaging in social activities, principally play and aggression. In one late-night interaction, for example, a sizable group of immature birds formed a screaming circle around a pair of keas that were involved in a particularly vicious brawl, much like gang members gathered around a knife fight. When the party finally winds down, the gang disperses from the site, and the birds roost in the forest. Keas that were radio-tracked in midwinter at Mount Cook National Park would sometimes leave the beech forest to roost at lower elevations. Once they went to roost, these birds seldom changed position between midevening and sunrise. When they did move at night, they flew only short distances.[44]

As this sketch of a day's activities suggests, dining in a kea group entails much more than just filling one's stomach. Group foraging presents, in a particularly concentrated form, all the diverse features of social life in the kea community, including aggression, dominance, courtship, and play. Young keas face a task of considerable magnitude: they must acquire expertise not just in foraging but in managing the social system as well. In the next chapter we examine how young keas bring about this transition to adult life.

FOUR

Growing and Learning

> We know that many young birds have, at the start, a very "open mind" with regard to food; they respond to an enormous variety of objects, edible and inedible alike, and learn to confine themselves to those they find edible. My suggestion is that we have as yet no more than the faintest idea of the kinds of things such birds learn when young.
>
> Niko Tinbergen, "On Aims and Methods of Ethology"

A typical kea nest contains two to three eggs. Once the eggs are laid, the female kea stays on the nest to incubate them, leaving for only brief intervals. She is fed by the male throughout the incubation period, which lasts three weeks. When they hatch, kea chicks are blind and utterly helpless. For the first month the male feeds the female, and she feeds the chicks. Gradually the male comes to feed the chicks directly, even while he is still feeding the female. The chicks develop very slowly, leaving the nest at between nine and thirteen weeks of age. By this time they are

This four-day-old kea was hatched at the San Diego Zoo. (Photograph by Judy Diamond)

being fed entirely by the male. Because of egg sterility and food shortages, usually only one or two young fledge from each nest.[1]

After they leave the nest, young keas develop through a series of distinctive stages. Fledglings remain completely dependent on their father for food; the male continues to regurgitate to them for at least an additional five to six weeks. Juveniles, or birds in their second summer, have dispersed from their natal ranges and travel from place to place in loosely organized flocks. Subadults have settled down after their dispersal and are generally two to three years old. Male keas breed for the first time at roughly four or five years of age; females may breed at as early as three years.[2]

When we began to compare the behavior of the different age classes to look for signs of maturation, what we found most remarkable was that the primary developmental changes we observed in the field were changes in social behavior, not in foraging. In a sense, what young keas learn most about when foraging with others is their social status and how that status can be manipulated to obtain access to food. They acquire foraging expertise within this larger context of their own social role. And because changes in this role are not simple or direct, social factors influence and constrain the birds' foraging behavior in different ways during successive developmental stages. Similar factors mark the course of human development. Toddlers are cared for directly and seldom allowed out of an adult's sight. Children are expected to be more independent, but they are also protected and given preferential access to resources. Adolescents receive less tolerance from adults and are in many ways expected to manage for themselves.

FLEDGLINGS

Fledglings at the study site spend a substantial amount of time searching, but they find relatively little on their own. They are limited to less effective foraging techniques, such as scraping meat from bones and gleaning particles of food overlooked by older birds.[3] When they attempt the more strenuous exploratory behaviors characteristic of adult males, they are too uncoordinated to execute them properly. For example, to tear open a bag, an older kea first grasps the bag firmly with its bill, then plants a foot against it for leverage, and finally pulls backward with the neck and back. Fledglings perform all three steps, but they often fail to get the sequence right. They pull a bit first, then stop and push the bag with one foot, then stand on both feet and pull again.

Not only do fledglings lack the strength and coordination for fully productive foraging, but they also seem almost completely unable to discriminate edible from inedible items. This inability may be a consequence of regurgitant feeding: the young birds have never been directly exposed to the feeling and appearance of unprocessed food. Their inexperience may not have been a disadvantage in the early evolution of the kea, since little in their natural habitat would injure or poison an inquisitive fledgling. In addition, young birds' attraction to adult feeding locations would ordinarily ensure that they eventually learned to recognize food.

A dump is not a benign environment, however. We have seen fledglings eating or attempting to eat paper tissues, rubber bands, string, masking tape, flashlight batteries, and foam rubber, among other unsuitable or dangerous materials. When trash was being burned, we even saw them attempt to bite the flames. Fiberglass batting is sufficiently attractive and lethal to young keas that unscrupulous run owners have occasionally strewn it about their property, hoping to poison the birds. The fledgling kea's open-minded attitude toward potential foodstuffs is not unique; other young birds also manipulate inappropriate objects.[4] But it exacerbates the enormous burden of individual learning faced by young keas and the hazards they encounter in exploiting human food resources.

Fledglings spend hours at the study site mauling and destroying inedible objects, in what is essentially continuous object play. In one season four fledglings spent the better part of a week ripping up a pile of old carpet pads. When some accommodating visitor left an overstuffed armchair at the edge of the dump, a mob of fledglings and juveniles had a riotous time, reducing it to springs and scarred wood over the course of several days.

Fledglings flock tightly with their nest mates and parents and are generally present at the site only when a parent is in the vicinity. There is no evidence, however, that they pay attention to the foraging of their parents or other experienced birds. The extensive social interactions of fledglings are indiscriminate with respect to foraging; the birds are no more likely to approach foraging than nonforaging individuals.[5] They interact primarily with other young keas, not adults. It is not even clear that fledglings recognize why their elders are attracted to the dump.

The social interactions of fledglings, except for requests for food or preening from their parents, are highly aggressive. Fledglings are far more likely to be attacked than any other age group and are also more likely to initiate an attack. Many of these aggressive interactions occur during the course of social play. For fledglings the boundaries between play and aggression are imprecise, and their interactions with other fledglings often slide back and forth between tussle play and serious violence. For older birds, locking bills and pushing with the feet are characteristic features of tussle play, but fledglings invariably include biting and other attack behaviors virtually never seen in the play of juveniles and subadults.[6] Young keas appear to emerge from the nest fully prepared to attack other birds indiscriminately, and only after receiving repeated drubbings and continuous harassment do they begin to learn discretion.

The most characteristic and conspicuous aggressive display in keas is the "wing-hold" posture, which even fledglings express

Fledgling and juvenile keas demolish an old chair left at the study site. (Photograph by Judy Diamond)

in fully developed form.[7] The body is held erect; the head feathers, particularly those on the nape and crown, are ruffled, forming a "hawk face"; and the body feathers are fluffed. The bill is held parallel to the ground or oriented slightly downward, and the tail feathers are fanned. The bird spreads its wings laterally, drooping them at the wrist, thereby exposing the scarlet underwing feathers, in a posture reminiscent of a heraldic eagle. To enhance the display the birds usually gape their bills and scream continuously.

Fledglings and juveniles often present the wing-hold display

88 / *Growing and Learning*

This young kea shows the wing-hold display, which indicates aggressiveness. In the most extreme form of this display, the wings would be fully extended. (Photograph by Judy Diamond)

in response to aggression and harassment by older individuals. But some birds characteristically adopt this posture immediately upon arrival at the site, even when there are no other birds in their immediate vicinity. The display appears to advertise a high level of aggressiveness, which often challenges other birds to attack. In our experience, a bird in the wing-hold posture will not readily attack others but will vigorously defend itself if attacked.

When they first emerge from the nest, males and females differ little in either social or foraging behavior. Female fledglings are not conspicuously less aggressive than males; they play as much and are, if anything, more destructive. And they are certainly as ineffective as males at discovering and exploiting new food resources. From this initial state of relative equality, however, the course of development in male and female keas diverges rapidly, and contrasts between the sexes are evident even in juveniles, only a year after fledging.[8]

JUVENILES

Juveniles are only a bit better than fledglings at excavating new resources and spend far less time scraping and gleaning. Nonetheless, they have substantially greater command of food resources than younger birds and invest less effort than other age groups in scouring the dump for food. Their high foraging effectiveness partly reflects a willingness to accept less desirable foodstuffs. Juveniles, like older birds, are attracted to foods that are high in sugar and fat, but adult males closely control access to such high-energy resources. As a result, juveniles at the dump mainly eat uncooked vegetables, fruits, and bread.

Juveniles are also more effective than other age groups at obtaining food by social means. They are quite assertive and do not hesitate to approach even the most dominant adults. With the aid of a distinctive display, termed "hunching," juveniles can often share an adult's food or even displace a feeding adult with only relatively minor consequences.

The hunch looks a bit like a formal Japanese bow, giving the bird the air of a subservient samurai before the shogun. The display blends the fledgling crouch posture with the aggressive wing-hold. In the solicitation crouch, a fledgling pulls its bill close to its breast. Hunching juveniles often exaggerate this movement, bowing their heads and drawing their bills in; even to the point that their heads touch the ground. They also fluff their

90 / *Growing and Learning*

A juvenile kea hunches to an adult. The hunch display gives juvenile keas preferential access to food. (Illustration by Mark Marcuson)

body feathers and extend their wings—not laterally, however, as in the wing-hold display, but rather down alongside the body, with the wingtips scraping the ground. Screaming and gaping are common in hunch postures, particularly in response to attacks.[9]

Although juveniles perform by far the most hunches, fledgling males sometimes adopt a less distinctive hunch posture: they draw down their wings, decline their bills, and ruffle their body feathers, but not to the extent that juveniles do. Hunches are seen less often in female than in male juveniles and almost never in subadults or adults of either sex.[10]

A juvenile assumes the hunch posture to approach an adult male and then stands directly in front of him, hunching and vocalizing. Often the juvenile comes so close to the adult that it effectively pushes him out of the way or shoulders him aside. By approaching adults in this posture the juvenile is roughly twice as likely to be tolerated and allowed to feed nearby. Furthermore, juveniles that hunch in response to an adult's approach are seldom displaced and are usually allowed to remain feeding.

Occasionally juveniles hunch to subadults and, less often, to other juveniles. Hunches to subadults occur about half as often as those to adults, and they are generally less successful. In contrast to fledglings, who solicit food only from their fathers, juveniles do not hunch exclusively to any one individual. We watched one juvenile hunch to three adults and two subadults in the course of an hour.[11]

Hunching may develop in a graded fashion as each young bird passes from fledgling to juvenile status. We speculate that as the frequency of direct feeding by adults declines, fledglings become more aggressive in their demands. They quickly learn that adults will retaliate if attacked directly but that the hunch posture elicits at least some residual tolerance, enabling them to obtain some of the resource. The posture itself reflects their mixed motivation. The approach is an aggressive act, an attempt to displace the adult from food. To fend off a responsive attack, the juvenile fluffs up and displays components of the aggressive wing-hold. But it reduces its value as a stimulus for attack by concealing the red underwings and lowering its bill in a posture reminiscent of a fledgling requesting food. Since the solicitation posture tempers aggression, the adult does not attack; meanwhile the defensive aggressive component of the behavior often displaces the adult from the food resource.

Because it reduces the likelihood of aggression by adults,

hunching may be considered a classic appeasement display, akin to the greeting ceremonies of seabirds. Appeasement carries the connotation of resisting or refraining from aggression; it is not, however, a sign of submissiveness. Indeed, hunching keas are commonly among the most dominant and aggressive juveniles in the group, and they perform the behavior while approaching feeding adults, not while attempting to avoid them. Submissive keas instead perform the crouch display. They are seldom attacked, whereas hunching birds are often at least pecked and buffeted around a bit. A hunching kea is stating that it will not attack, but it will defend itself if attacked. A crouching kea, in contrast, indicates that it is not even prepared to defend itself and is predisposed to flee.[12]

The fledglings seen at the study site had probably emerged from nests in the immediate area, but the juveniles appear to have come from a much larger region, since there were many more of them than could have been produced locally in the previous season's nesting effort. Keas spend the first year after leaving the nest simply wandering through the mountains, often in large flocks. Some of these birds disperse great distances from their natal ranges. Juvenile keas have occasionally been reported in bird surveys of Christchurch Airport, nearly 80 km from the nearest suitable kea habitat. Especially during fall and winter, juveniles collect where food is easily accessible even for animals that lack foraging experience, such as at dumps, cabins, and parking lots. The New Zealand term for a large, loosely organized group of animals is a "mob," most apt for a flock of juvenile keas. Keas are fairly destructive at all ages, but the worst of this devastation is inflicted by mobs of idle juveniles.[13]

SUBADULTS

The largest change in social behavior comes in the transition from juvenile to subadult status. Like human teenagers, subadult keas undergo an adolescence, a massive change in their social status, their maturity, and their ability to survive on their own. Some of the behavior of subadults foreshadows the roles they will have as adults. Play offers the clearest example. Juveniles and fledglings engage primarily in tussle play and object manipulation. But like adult males, subadult males rarely engage in play of any sort, and when they do, the episodes tend to be brief. Subadult females, on the other hand, engage with males in extensive bouts of both tussle and toss play, as do adult females. Their play thus begins to suggest elements of courtship.

Subadult keas receive neither the benefits of adult expertise nor the comforts of childhood. They eat less than adults, even though they spend as much time as birds of other age groups searching for food. Subadult females, in particular, spend almost as much time excavating as adult males. Subadults are most successful at finding food during the afternoons, when there are virtually no adults at the study site.[14]

Subadults rarely exhibit the hunch displays that juveniles use to obtain access to food found by adults, and when they do, the display does not appear to work. Lacking the protected status of juveniles, subadults are readily displaced from food resources by adults and are unable to displace the older birds. They commonly crouch in interaction with adults; they are more often the recipients than the instigators of aggressive interactions with adults, and they seldom emerge as victors in these encounters. Paradoxically, given their subordinate status in interactions with adults, subadults generate much of the violence and harassment inflicted on fledglings, and they will often attack juveniles, seemingly undissuaded by hunches or wing-holds.[15]

To compensate for their relative hardships, subadults use one foraging strategy more effectively than any other age group of keas: they often seek out keas that are feeding and attempt to steal their food. They begin by stealthily approaching a feeding bird, usually in a submissive crouch. They watch intently, and when their victim is momentarily distracted, they grab the food item and fly away with it. Food piracy is hardly unique to keas. Many seabirds, particularly gulls, are master thieves. Such thievery has, however, never been recorded in the wild in any other species of parrot.[16]

In keas the transition between juvenile and subadult is associated with physiological changes. Subadults have molted their ragged juvenile plumage, worn since they initially emerged from the nesting cavity. The changes that bring on the molt also affect the remaining yellow coloration in the cere and lower bill, transforming the bird's appearance to that of a young adult. The transformation generally occurs earlier in females than in males. Many females banded as fledglings or juveniles and seen at the site in successive field seasons apparently passed through their subadult stage to full adulthood within a year. Male juveniles, in contrast, were invariably still subadults when they were re-sighted. Our observations suggest that males remain subadults for at least two years. This sex difference in transformation to the adult color pattern corresponds to the earlier onset of sexual maturity in females. Jackson notes that female keas often begin breeding in their third year, while males do not breed until they are at least four or five years of age.[17]

What happens to juvenile birds that initiates the profound change in their behavior, status, and morphology? Possibly the deciding factor is dispersal. Between the time of fledging and the summer when they become subadults, keas disperse from the areas in which they hatched. They wander the mountains with other juveniles, effectively released from the social structure provided by the local adult population. This release from social inhibitions may well be what initiates the physiological change that gives them their mature physical appearance and social status.[18] Once they have lost their yellow bills and eye rings, they cease their wandering and settle down in one place. Subadult females fairly often return to the area where they were hatched, but very few of the male fledglings and juveniles we banded were ever seen again. It seems likely that many of them settled down in neighboring regions. If so, that might explain the anomalously low status of subadult males. As strangers in the area, they would occupy the bottom of the feeding hierarchy and would have to work their way up by aggressive interactions with established individuals.

THE ROLE OF LEARNING IN KEA SOCIETY

Not all birds have to learn as much as keas. Many hatch with an inherent knowledge of much of what they will need to know to survive and reproduce, including what is edible and what is not. Great crested grebes, for example, do not need to learn food-related stimuli; indeed, in some instances they cannot learn them. Their foraging behavior is so constrained that only the movement of live fish elicits prey catching and feeding, even if the birds are hungry and fresh dead fish are available.[19]

In other birds, however, behavior is less predetermined. A few generalized techniques are programmed in these birds, and they are applied with great versatility to a wide range of circumstances. A classic example is the raven, which appears to have relatively little that is fixed about its nature at birth and to require a great deal of learning in the course of its development.[20] The same can be said of keas. Fledgling keas emerge from the nest in a state of complete naïveté. They do not distinguish edible from inedible items and show no fear of dangerous animals or situations. The foraging actions that develop earliest are pulling and prying at objects, but they do not discriminate among the objects they manipulate. Moreover, these early behaviors are not integrated into the coherent patterns that are necessary for successful foraging.

The process by which their primordial ignorance is transformed into the extraordinary richness and complexity of adult foraging and social behavior is perhaps the most fascinating aspect of kea biology. This process is by no means random or arbitrary. Young keas have two essential predispositions that shape all their subsequent learning. First, they are strongly socially facilitated—attracted to other keas and fascinated by the activities they are engaged in. Second, they are almost compulsively motivated to play with other keas and with objects in their environment.

Social facilitation, narrowly defined, is the increased performance of a behavior in the presence of another individual performing that behavior. This narrow definition, however, is often difficult to apply to foraging situations. Other individuals may be performing a sequence of actions (e.g., pulling, prying, rummaging, eating) rather than a single behavior, and furthermore the actions are not performed in isolation but are addressed to particular objects or food items. We prefer to use the term in a broader sense that includes an increased responsiveness to particular objects as a result of another's interaction with them. Thus, social facilitation ensures that a kea will pay attention not just to what other keas are doing but also to what they are feeding on or interacting with.[21]

Social facilitation seems to account for most of the influence exerted by other members of the group on learning in young keas. Juvenile keas do not appear to learn novel foraging techniques by imitation. When they do use a technique they have observed, it is generally a simple movement, such as overturning a rock, that was well within their competence and that they had probably often performed before. Conversely, we could find no evidence that adult keas modify their own behavior to enhance their young's learning, as might be expected if keas engaged in teaching. Nor do adult keas make any effort to protect their young from the consequences of their actions. They do not, for example, attempt to prevent fledglings from burning themselves in dump fires, even though the adults clearly know the flames are dangerous and actively avoid being scorched themselves. Like many social primates, adult keas seem remarkably indifferent to the hazards faced by their offspring.[22]

Even within the limits of social facilitation, however, young keas do not make use of all the information supplied by their elders. Young keas may learn the appearance of food and its location through observing others, but they do not perform the particular actions of the foraging adult. Juveniles and fledglings are drawn to the objects of adult interest, but once in possession of such an item, they appear to approach the task of dismantling and consuming it without socially mediated preconceptions. They are guided more by the affordances of the object than by the behavior of other individuals. Soft things are pierced or pulled on, hard ones are scraped, holes and crevices are probed and pried. Beyond this, they probably rely on individual trial-and-error learning, sampling from their repertoire of foraging actions to see what works on a given type of food, and eventually coming across an efficient means of handling it.[23]

Social interaction may impede, as well as enhance, individual trial-and-error learning. The dominance hierarchy reduces the opportunities for individual foraging among young keas. A juvenile or subadult that finds a choice food item on its own has little chance of retaining it. Not even a hunch display would prevent a hungry adult from displacing a juvenile from a valued resource. So although social facilitation helps young keas recognize and locate food, the social structure reduces the incentive and opportunity for them to practice individual foraging.[24]

The role of social facilitation changes as keas mature. Fledglings begin independent life manipulating and demolishing anything within reach. Although they interact sparingly with adult keas, they often follow along after their parents and occasionally pick up foods sampled or discarded by them. Juveniles are much more socially oriented and frequently approach even unrelated adults, from whom they may obtain indirect access to food. But juveniles will as readily approach a bird that is not eating as one that is. Subadults, by contrast, are highly selective in their social interactions, approaching mainly individuals that are in possession of valuable resources. The development of foraging can thus be seen as a transition from relatively nonsocial and indiscriminate exploring to focused social interaction with adults; attention then shifts to the objects that adults control and finally to the value and utility of these items.

The second predisposition to learning in keas is play. By imparting broad strategies for dealing with novel situations, play ensures that young keas will be able to adjust to new items in their environment. In the course of play, young keas may become more familiar with the properties of objects and with the nature of their own relationships to other members of the social group, but this learning appears to be largely incidental. The kea's persistence at repetitive play interactions suggests that the acquisition of specific information is not their primary function.

We believe that the primary function of play for the kea is to maintain behavioral flexibility—the animal's capacity to modify its behavior or ecology when faced with new challenges, in Robert Fagen's definition. Fagen, a play researcher, asserts that play improves behavioral flexibility by producing "complex generalized skills for interacting with varied and novel social and physical environments." He speculates that play stimulates the growth and proliferation of neurons in the brain, enhancing the animal's ability to respond to novelty. Valerius Geist proposed a similar concept, contending that play induces a variety of changes in brain anatomy and chemistry that increase the adaptability or versatility of behavior. In this model the influence of play on learning is indirect: play does not necessarily yield information and experience that an animal requires for particular environmental circumstances, but it engenders a neurological readiness for adaptive response.[25]

Like facilitation, play in keas moves through successive developmental stages. Fledgling play is chaotic, with little to differentiate it from aggression. Juvenile play, however, is firmly established as a distinct activity, separate from other kinds of social behavior. It takes on the characteristic structure of tussle play, with very high levels of facilitated behaviors between playing partners. In juveniles play also begins to be differentiated between the sexes, with males showing much higher levels of play than females. The differentiation continues in subadults: females adopt the typical adult female pattern of toss play with adult and subadult males and tussle play with younger males, while subadult males begin to display the behavior patterns of adult males, greatly decreasing their levels of play.

Together the two predispositions to learning provide young keas with essential equipment. The combination of strong social facilitation and individual motivation to play with other keas and with objects brings both conservative and innovative elements to the kea's personality. Through facilitation, young keas benefit from the knowledge possessed by older individuals. At the same time, their play ensures they will retain the flexibility needed to adapt to constantly changing conditions.

The present-day learning strategies of keas evolved during their isolation among the glaciers of the South Island. The ecology of that environment left an irrevocable stamp on the kea's personality, making the kea distinct even from its closest relative, the kaka of the lowland rainforests. The contrast between these two species sheds light on why many features of the kea's personality are found in no other New Zealand parrot.

FIVE

The Prince and the Pauper

> Comparative ethology can provide a description of what must have happened in the past. In addition, the study of survival value or function allows us to formulate at least likely guesses about the way the evolution of a species has been controlled.
> Niko Tinbergen, "The Evolution of Behavior in Gulls"

Unlike bones or teeth or footprints, an animal's behavior doesn't fossilize. We can determine little of the evolutionary origins of the kea's mode of life from fragmentary remains among the debris of ancient caves. The study of behavioral evolution necessarily employs comparative methods, similar to those Charles Darwin used in his original investigations of the mechanism of evolution. By contrasting the behavior patterns and life history of one species with those of closely related species, differences can be noted, yielding a narrative account of how and why the behavior may have changed. Although comparative data can never firmly establish cause and effect, the evidence can be very compelling. Differences in the behavior of closely related species can point to some of the ecological factors that may have contributed to its evolution.[1]

The kaka, which inhabits the lowland forests of all three of the main islands of New Zealand, is the kea's closest living relative. The two species belong to the same genus and are in effect cousins in a lineage that is otherwise wholly distinctive, with no clear relationship to other groups. Their kinship is readily visible. Keas and kakas are both medium-sized parrots with subtle color patterns. Where keas are olive green, kakas are olive brown, with gray crown feathers and a purplish red breast and abdomen. Kakas are generally a bit smaller than keas. In the summer healthy male keas weigh in at about a kilogram. The North Island kaka subspecies weighs less than half of that, roughly 400–500 g; South Island kakas average closer to 700 g. The skeletal structure of the two species is so similar that until recently naturalists were uncertain how to identify the fossil remains of *Nestor* parrots.[2]

The most conspicuous differences in the appearance of keas and kakas are in their upper bills, or culmens. In kakas the culmen is more typically parrotlike, deep and robust, suited to crushing and gouging. In keas the culmen is slender and hooked, suited to prying and piercing. In both keas and kakas, males are larger than females and have a much longer culmen. The degree of these sexual differences is virtually identical in the two species. In addition to the contrast in bill shape, the bones of the legs and wings are significantly longer in keas than in kakas, which presumably reflects the kea's adaptation to foraging on the ground. Keas are primarily, though not exclusively, ground feeders. Their relatively long legs enable them to walk smoothly, and

A North Island kaka on Kapiti Island. (Photograph by Judy Diamond)

Comparison of the bills of the kea *(left)* and kaka *(right)*. (Illustration by Mark Marcuson)

they can also run at a surprising clip, using an odd skipping step that alternates striding and hopping. Kakas, in contrast, spend a far greater portion of their waking hours in trees. When they do descend to the ground, their gait is relatively slow and awkward. In the trees, however, kakas are as active as monkeys, often hanging by their bills or dangling upside-down by one or both feet.[3]

The kaka's color pattern, like the kea's, changes with maturity, but in kakas the patterns are less conspicuous. Fledgling keas and kakas both have yellow eye rings, yellow ceres, and yellow in the corners of their mouths. In keas these areas are a bright yellow-orange, visible from long distances; in kakas there is only a pale yellow tinge. The yellow markings in kakas fade after several months, when they cease begging from their parents and become juveniles, but the eye ring appears pale white until the bird is about a year old. After that point a kaka's age can no longer be determined on the basis of markings. The kea's color pattern indicates its age for at least three years, by which time the cere, eye ring, mouth, and bill have all turned dark brown. In most kakas the juvenile coloration turns to black after their first year.[4] On Kapiti Island, however, and possibly elsewhere, adult female kakas retain the white eye ring of juveniles.

The geographical distribution of keas and kakas shows surprising correspondences. Keas have the highest alpine regions pretty much to themselves. The ranges of the two species overlap, however, at moderate altitudes in wet forests in many areas of the South Island, including northwest Nelson, Fiordland, and south Westland. Both keas and kakas are found throughout the forests of the southwest coast of the South Island. The two species show comparable habitat preferences, spending more of their foraging time on beech trees than on any other species of tree. Keas are never as common as kakas in lowland forest, however, and even there they are found mainly at higher elevations.[5]

The dietary preferences of keas and kakas also overlap. Kakas, like keas, are fond of high-energy food resources, such as nectar and honeydew. On the South Island, kakas spend up to half their time feeding on the sugary sap from forest trees. In beech forest in the springtime, kakas collect honeydew from scale insects and also consume a fair number of the insects themselves; in the winter they lick sap directly from the trees, either by stripping the bark from the trunk or by gouging holes through the bark to the cambium. When the southern rata blooms in summer in the broadleaved rainforests on the west coast of the South Island, kakas spend much of their time drinking the nectar. Keas are also attracted to this resource and will descend to lower elevations to gorge themselves at rata trees; they have even been observed feeding with kakas in the same tree.[6]

The kaka's preference for foraging in trees accounts for many of the dietary differences between the species. But the kaka's thicker, more robust upper bill probably also makes some foods more readily accessible to kakas than to keas. Both species consume insects, for example, but keas do so by tearing apart rotten logs for grubs and foraging among tussocks for grasshoppers. Kakas, by contrast, are able to chisel beetle larvae and other insects out of standing trees. In the mountain beech forests of the South Island, kakas spend up to a third of their foraging time extracting fat beetle larvae from tree trunks. They also seek out forest insects in the broad-leaved rainforests on the South Island and on Kapiti Island. Both keas and kakas consume large amounts of fruit when available, but kakas also extract and crush nuts and seeds. Keas eat very few seeds, probably because their bills are not as well suited to crushing and grinding.[7]

The biggest dietary difference between keas and kakas lies in their attitude toward carrion. Kakas have never been observed to feed at carcasses and have never been recorded to kill and eat other vertebrates, even in captivity. They have no history of harassing sheep or other livestock and have been considered pests only for their depredation of orchards.[8]

OBSERVING KAKAS ON KAPITI ISLAND

To provide a basis for our comparisons between keas and kakas, we conducted a field study of kakas on Kapiti Island, a bird refuge off the southern coast of the North Island. Kapiti is a large island of over 2,000 hectares, separated from the mainland by about 6 km of ocean. Rainfall is moderate, with a yearly accumulation of a bit over a meter, most of it falling during the winter. Temperatures are generally cool but are far more equable than those of the kea's habitat in the Southern Alps.[9]

Kapiti is an optimal location for kaka research. Its ridges and hollows are cloaked in a thick lowland forest, the habitat for about a thousand of the birds. Although kakas were once common in the temperate rainforests of both the North and South Islands, their numbers have declined dramatically since European colonization. On the South Island, kakas still hold their own in indigenous forest along the southwest coast, but they are absent from most areas of the North Island. Classified as a vulnerable species, kakas face high risk of extinction in the wild in the medium-term future. The kakas on Kapiti now constitute one of the largest existing populations of the North Island subspecies.[10]

Kakas share the island's forests with an abundance of other native birds, including North Island robins, tuis, bellbirds, parakeets, pigeons, and cuckoos. Kapiti is also a refuge for some of New Zealand's rarest birds. Five little spotted kiwis were moved to the island in 1912 and eventually established a stable breeding population there. This may be the last viable population of the species, which has not been seen on the mainland since 1952. Populations of stitchbirds and North Island saddlebacks have been introduced to Kapiti and appear to be thriving; neither species has been seen on the mainland since the turn of the century. Several pairs of highly endangered takahes, New Zealand's huge flightless gallinule, were recently brought to the island for protection and now also breed there.[11]

This rich and exotic habitat is all the more remarkable in the light of the island's history. Once a Maori stronghold, Kapiti served as a whaling station in the early 1800s. After the whaling industry went into decline, sheep, cattle, and goats were introduced to the island, along with brush-tailed possums, cats, and rats. Fully half of the original forest acreage was cleared for grazing, and large parts of the rest were damaged in a series of massive fires. The potential value of the island as a nature preserve was recognized at the turn of the century, however, and efforts were begun to reestablish the native flora and to eliminate introduced species. The free-ranging cattle, goats, and cats had been eliminated by the 1930s, sheep grazing was halted in 1969, and a concerted, systematic campaign of poisoning, trapping, and hunting with dogs finally eliminated the last possums in 1986. These strenuous efforts were rewarded by the regrowth of a forest which in places now contains a greater diversity of plant species than any comparable area on the mainland.[12]

In 1990 and 1991 we obtained permits to visit Kapiti Island to observe the behavior of kakas in the field and record their vocalizations. The only permanent year-round resident of the island is a ranger from the New Zealand Department of Conservation. To collect visitors, he crosses the channel toward the sandy beaches on the mainland but stops his boat within about 50 m of the shore. To avoid the possibility of unintentionally transporting black rats to the island, his passengers have to wade out through rough surf, carrying their baggage high above their heads.

While conducting our research we stayed in a whare, or cabin, that was built of native totara wood in 1897 by Kapiti's first caretaker, Malcolm McLean.[13] We immediately recognized one of the most conspicuous differences between keas and kakas: kakas are far less inquisitive and destructive. The doors and windows of the whare are commonly left open during the day, even when no one is there. The open door was an invitation to takahes, who, ignoring our presence, boldly marched around the kitchen like snoopy landlords. The kakas, however, never ventured farther than the doorsill, waiting patiently for us to notice them and provide a small treat. Similar cabins in the mountains of the South Island must be shuttered and latched when unoccupied to avoid complete demolition by roving gangs of juvenile keas.

Kakas are readily observed at close quarters on Kapiti. Since the early 1960s the resident ranger has put out supplemental food—a solution of brown sugar and molasses in water—four times a day at what has become an established group-feeding site, a metal-lined wooden water trough about 2 m long and 30 cm wide. Kakas gather in the trees around the site at feeding time, competing to gain access to the sugar water. About 4 m from the trough we set up a small canvas blind, which served to keep us dry and to keep the birds away from our computer equipment. Kakas were the primary visitors at the site, although native honey-eaters such as tuis and bellbirds also came to feed on the sugar water.

Ron Moorhouse, a researcher from Victoria University in Wellington, had been studying kaka ecology and reproduction on Kapiti Island during the previous three years. In the course of his studies he banded more than thirty birds, fifteen of which were regular visitors at the feeding trough during our study. He provided detailed information about the sexes and ages of these birds, enabling us to contrast the foraging and social behavior of particular sex and age groups. Moorhouse's work remains the most thorough study ever conducted on the North Island kaka subspecies. Together with field work on the South Island subspecies by Jacqueline Beggs, Peter Wilson, and Colin O'Donnell, it gave us an additional basis for comparison of the ecology and life history of keas and kakas.[14]

The feeding trough on Kapiti created a social context fully comparable to that of the kea study site in Arthur's Pass National Park. In both places the birds were foraging at a traditional location, and the concentration of resources assured a high level of social interaction. Moreover, the local parrot populations that gathered at the two sites were surprisingly similar. Using data from resighting of known individuals, we estimated that sixty-six keas made use of the Arthur's Pass study site; applying the same technique on Kapiti, we judged that fifty-two kakas were regular visitors at the feeding trough. About 28 percent of the kaka aggregation were females, however, in contrast to only 11 percent of the kea group.[15]

BEHAVIORAL CONTRASTS

To a kea researcher, the most immediately noticeable feature of kaka behavior is the birds' relative geniality. Although the overall frequency of aggressive interactions at the Kapiti feeding trough was comparable to that at the Arthur's Pass study site, the intensity of the Kapiti interactions was much lower. Kakas are far less aggressive than keas, even while foraging at close quarters in large groups. They usually resolve conflicts by displays rather than combat. When confronted by competitors, kakas threaten each other with "stare-down" displays similar to those shown by adult male keas. Beyond that, they might gape their bills at each other, fan out their tail feathers, and occasionally push each other with their feet. Full-scale battles, involving biting or bludgeoning each other with their wings, are exceedingly rare.

Kaka fledglings become independent of adults within six months, fully a year before keas reach a similar stage. As juveniles kakas are neither fed nor deferred to by adults. They evidently lack the preferred position that juveniles attain in kea society: in foraging, they cannot displace older, more dominant male or female birds and show no characteristic display to deter adult aggression. This contrast with the status of juvenile keas may be related to the relative inconspicuousness and rapid fading of the kaka's juvenile coloration. Given the absence of a persistent, distinctive juvenile plumage, it is scarcely surprising that adult kakas do not differentiate juveniles from older individuals.[16]

In kaka society adult females, rather than juvenile males, use a specialized display to gain access to resources. Nesting female kakas show a "pout" display, similar in function to the hunch posture of juvenile keas. They hold out their wings slightly and draw them down, exposing the bright red color beneath, while ruffling their head feathers and inflating their breasts. They also emit a continuous, grating screech, emphasized by vigorous head-bobbing. Jackson remarked that the displaying female kaka sounds as if she were "snorting through her nose."[17]

This rather alarming behavior not only allows the displaying bird to displace other animals from rich foods, but it also acts as a defense against dominant male kakas, who seem unable to drive displaying females away. Food solicitation may, however, be the primary function of the display: pouting females are invariably actively breeding birds, and the display commonly precedes feeding of the female by her mate. Thus, a display that helps to ensure that juvenile keas will get enough food in their stressful alpine environment has an analogue in kaka society that serves to provide resources to nesting females.

We occasionally observed keas performing a display identical in form and similar in function to the kaka's pout. Once a juvenile male kea had been scraping dried meat from a piece of bone when an adult male launched a vigorous and unprovoked attack. The juvenile screamed repeatedly and adopted an extreme hunch posture, almost standing on his head. Moments later he lifted his head and began to bob it rapidly while making a guttural rasping vocalization, exactly as female kakas do during pouting. The similarity in these displays suggests that the action pattern of pouting may be evolutionarily ancient, originating in the proto-kaka, the species from which both keas and kakas evolved.

The courtship behavior of keas and kakas is wholly disparate. Adult male kakas often court females in the area of the feeding trough, sidling up to them with a warbling vocalization. The courting male rapidly alternates his orientation, turning first away from the female and then toward her. When he comes within about 20 cm of her he crouches and raises his wing, exposing the red underwing.[18] Among the kakas on Kapiti we saw none of the mutual jumping and flapping and the tossing of objects that appears to characterize courtship in keas.

The vocal repertoires of the two species are also quite distinctive. Keas produce a number of variants on the basic *kee-ah* vocalization, ranging from the soft *meows* of birds preparing for sleep to the "bleat-trills" that signal imminent flight. They also make several different warning sounds and a variety of squealing and gurgling noises that are characteristic of play or courtship.

The pout display of the female kaka *(left)* resembles the hunch and wing-hold displays of young keas. An adult male kaka *(below)* feeds a nesting female in response to her pout display. (Photographs by Judy Diamond)

Kaka calls, however, range from whistles, twitters, and chirps to growls and raucous squawks and shrieks. Their vocalizations are far more diverse than those of even the most accomplished kea; in fact the kaka may have the largest vocal repertoire of any New Zealand bird.[19]

But the most salient contrast between the two species lies in play. Kakas do show solitary object play, in which they growl, hang upside-down, and tear apart sticks and branches, but such episodes are briefer, less varied in content, and considerably less common than comparable ones in keas.[20] Moreover, the capricious and rambunctious social play of keas is largely missing among their reserved cousins from the lowland forests. We have seen little social play by kakas of either sex or any age group. Like keas, kakas often linger in the neighboring trees after foraging. Allopreening, courtship, and other social behaviors are every bit as common among kakas in these circumstances as among their alpine counterparts. The absence of kea-style play, along with the lower levels of aggression in kaka foraging aggregations and the lack of a preferential status for juveniles, thus encapsulates the fundamental behavioral difference between the species.

ECOLOGICAL CONTRASTS

To survive in the alpine environment, the kea has to make use of a very broad range of resources. The foods available year-round in this habitat, such as beech leaves and daisy roots, are nutritionally incomplete. Bountiful years are rare and unpredictable, whereas seasons of dearth are frequent. The kaka's rainforest habitat, in contrast, provides more abundant and reliable food resources. The wet lowlands of New Zealand that are the kaka's primary home are far richer in fruit, insects, and nectar than the kea's mountain beech forest.[21]

In spite of the differences in their habitats, the time course of reproduction appears to be similar in the two species. Moorhouse studied the reproduction of kakas on Kapiti over a five-year period, obtaining data on fifty-one kaka nests. He discovered that kaka chicks hatch after a three-week incubation period and leave the nest in their tenth week after hatching. They are subsequently dependent on their parents for food for at least five or six weeks. From studying a set of thirty-six nests, Jackson reported a virtually identical development schedule for keas. Both species must thus begin egg-laying about three months before the time of peak food availability to ensure maximum survival of newly fledged young. Their contrasting environments thereby impose a striking difference in the scheduling of reproduction. Kakas lay their eggs mainly in November, in the mild spring of the lowland forests, and fledge their young in late summer, around the end of January. Summer does not last as long in the mountains, however, so keas are obliged to get their chicks out of the nest earlier. Some keas are fledged as early as October or November, with a substantial peak in December. To obtain this early fledging, keas must begin incubation sooner. Jackson observed some keas nesting as early as July and August, incubating their eggs under the snow in the frigid depths of the alpine winter.[22]

The reproductive rate of kakas per nesting attempt also appears to be similar to that of keas. Although kakas almost always lay four eggs, the incidence of sterile eggs is fairly high, and generally only two or three actually hatch. One or two of these kaka nestlings survive, on the average, the others having been killed by predators. In comparison, keas usually lay fewer eggs—two to three on average—but the rate of hatching and fledging is somewhat higher. The result is about one to two young produced per nest, much the same as for kakas.[23]

In spite of their similarities in reproduction, the aboriginal population level of kakas was almost certainly higher than that of keas, owing to the greater abundance of rainforest foods, and this differential is still apparent under protected conditions today. Modern kaka populations have been reduced on the main islands, mostly as a consequence of deforestation, predation, and competition from introduced species. Where kakas have been protected, however, they achieve densities greater than those ever attained by keas. The Kapiti Island nature preserve harbors about one kaka for every 2 hectares. Since introduced nest predators still remain on the island, even this population may not reflect the birds' abundance in precolonial New Zealand. In contrast, Arthur's Pass National Park sustains at most one kea per 32 hectares.[24] Given that nesting success is similar in keas and kakas, the alpine environment must be imposing a limit on other aspects of kea population dynamics. Either the mortality of the young after they leave the nest is higher in keas than in kakas or the proportion of the population that is able to breed in any given season is lower. It seems likely that both factors are operating.

Before predatory mammals such as rats, cats, and stoats were introduced to New Zealand, fledgling kakas emerged from their nest cavities into a fairly benign and predictable world. Food was reliably available and often abundant, much of it in the form of insects, fruit, and nectar, which did not require extensive experience to learn to exploit. Once a young kaka made it through to fledging, it had a rather good chance of survival. This comfortable situation still apparently obtains for young kakas on

Kapiti Island. Of the twelve juvenile kakas that Moorhouse had banded near the feeding trough between 1988 and 1991, all were still alive and still visiting the trough in 1995.[25]

Fledgling keas, on the other hand, face more forbidding prospects. As might be expected of birds from generally cool habitats, both keas and kakas have unusually high basal metabolic rates, maintaining an internal temperature of nearly 40°C. This condition greatly increases their need for high-energy foodstuffs, which compounds the challenges posed by the kea's limited food supply. Keas are larger than kakas, so even at the same ambient temperature they would need to consume twice as many calories as a North Island kaka. But in fact the Southern Alps are considerably colder than the lowland forests. On a winter day in Arthur's Pass National Park, the average temperature might be 1°C, while on Kapiti Island it would be closer to 9°C. Even though keas are well insulated against the cold, it is likely that their daily caloric requirements increase considerably in the wintertime.[26]

Our observations and those of Jackson suggest that a substantial proportion of juvenile keas—perhaps as many as two-thirds—fail to survive the winter months. Most of these young birds simply starve to death. In the winter even adult keas may go hungry for long periods or subsist mainly on minimally nutritious provender, such as fibrous stems and leaves. Jackson commonly found dead keas in winter whose stomachs contained nothing but green bile. Summer is not necessarily a propitious season either. Jackson describes in graphic detail the summer of 1957–58 in Arthur's Pass, when the rainfall was almost double the norm for that time of year. Many of the alpine shrubs failed to bloom, which resulted in a dearth of nectar and berries, and the abundance of insects was greatly reduced. Kea chicks died in their nests, and kea parents ate their eggs. The kea population of Arthur's Pass dropped by one-third that summer.[27]

Besides having greater juvenile mortality than kakas, keas probably breed less frequently. In most summers at Arthur's Pass, between 10 and 20 percent of adult male keas can be seen feeding fledglings, which suggests that the number of productive nests is probably quite low. Jackson noted that a female kea generally spends several years building her nest and establishing her claim to it before actually laying eggs. As a result, although many female keas build nests, only a fraction of the nests contain eggs in any given year. Among the kakas on Kapiti Island, in contrast, roughly 40 percent of the population attempts breeding in a typical year, and up to 80 percent of the birds may breed in a "boom" year.[28]

Why might breeding be more limited in keas than in kakas? An obvious reason is that there is usually not enough food available, even in summer, for large numbers of keas to raise young. Another possibility, however, is that nest sites may be limited, and there may not be enough for every pair of adults to set up housekeeping. Kakas invariably nest in tree cavities, which are not in critically short supply in the lowland forests, but keas appear to have greater limitations. Jackson found that most kea nests are in burrows under boulders on warm, north-facing slopes. The nests are all below tree line in the thickest part of the mountain beech forest, and the birds seem to prefer boulders large enough to emerge above the surrounding brush. These highly specific requirements may be related to the need for a warm, dry, and secure location for incubating eggs during the alpine winter. Nest sites that can ensure that the nestlings will survive to the coming of spring must be fairly rare. In fact Jackson saw considerable evidence that such sites were reused for generations, and favored perch locations outside the entrances were often thickly encrusted with kea droppings.[29]

Given the limitations on food and nest sites, kea society apparently evolved rather ruthless ways of determining which birds would be allowed to breed. Their rigid social hierarchy may limit sufficient access to food, and thereby reproductive opportunities, to only fairly dominant adult males. Adult males fight fiercely for social position, and Jackson noted that they often break up pairs of younger birds and drive them away from nest sites. Quantitative data directly relating dominance status to breeding are fairly meager, however. Over the course of our field study we observed four banded males with recently fledged offspring. Three of these males engaged in enough social interactions with other adults that we were able to gauge their level of dominance. All three ranked in the upper half of kea society, although none was among the most dominant birds in the group. Adult females may also play a role in restricting reproduction. They generally establish their own nest sites and build the nests themselves. Consistently aggressive to younger females, they actively exclude them from nest sites.[30]

Keas other than the breeding pair visit nests frequently and may even enter the nest burrow. The attitude of the resident pair toward these visitors, which may be of any age and sex, is quite variable. Sometimes they appear to tolerate the intrusion; sometimes they drive the visitors away. Since there is no evidence that keas are cooperative breeders, with birds other than the breeding pair helping to rear the young, the function of nest visiting is somewhat obscure. Visiting is not a wholly benign activity, however, and its consequences can occasionally be grim. Jackson observed five nests in which the eggs had been destroyed by keas and one nest in which week-old chicks had been killed by visitors and "trodden into the floor of the nest."[31] We surmise that at least some of these visitors may be attempting to break up a nesting effort by lower-status individuals.

THE SIGNIFICANCE OF PLAY

The impoverished food supply and limited nest sites in the alpine environment have made reproduction a rare privilege for keas, with social restrictions on who will be allowed to breed. This situation has apparently led to a delay in the age of first reproduction. Male keas seem unable to raise young until they are at least four or five years of age, and even among adults, only older, more dominant birds may be afforded the opportunity. Male kakas, on the other hand, reach sexual maturity a year earlier than keas, and no instances of interference with their nesting have been recorded.[32] Limited reproduction thus extends the juvenile period for keas, as a prolonged stage in which the young birds only gradually establish their claim to the privileges of adulthood.

Winter survival of juveniles, however, is chancy. Without some provision for tolerance from the otherwise rigid and aggressive adults, they would have little likelihood of living long enough to breed. Lenience toward juveniles, then, is a second logical outcome of the severity of the kea's habitat. Although adults cease to regurgitate to fledglings after a couple of months, they give way to juveniles in feeding aggregations. Since these young birds still have a rather limited ability to discover food on their own, this behavior undoubtedly improves their survival in the following winter.

The combination of delayed maturity and parental lenience has provided keas with a unique evolutionary opportunity. They enjoy an extended grace period early in their development, in which they are protected, to some degree, from the consequences of their lack of foraging expertise. One could say that they have been given a childhood, complete with its primary fringe benefit: the provision for time to play. A childhood only makes possible the evolution of play, however; it does not compel it. What, then, led the kea to adopt play as a pervasive element in its behavior?

Food is not only limited in the kea's alpine environment, but it is also unpredictable. Alternative food sources are transitory in the high mountains, with their abundance constantly changing even over the course of a few weeks. The richest resources are among the least reliable: nectar and pollen can disappear after a few weeks of heavy rains, carcasses are rare to begin with, and even beechnuts are only available in quantity during infrequent bonanza seasons. Keas evolved under circumstances in which they had to be ready to shift to new food resources on short notice.[33]

The kaka's resources, on the other hand, are less subject to fluctuations in availability. Winters in kaka forests are cool and wet, but there is no permanent snow, and a substantial amount of food remains accessible year-round. On Kapiti Island one or another of the native tree species is bearing fruit at almost any given time, there are two long seasons of nectar production, and arboreal insects can be captured all year. In the mixed beech and podocarp forests on the South Island, the mainstay of the kaka's diet is tree sap and honeydew, items that are readily available in any season.[34]

The disparity in the predictability of their environments molded the evolution of the kea and the kaka in different directions. The kea's changeable food resources selected for great flexibility in the use of a limited set of all-purpose foraging techniques. When presented with a novel object, the kea pulls out its metaphorical Swiss army knife and tries one blade after another. Its generalist ecology is thus expressed not just in the spectrum of foodstuffs it accepts but also in the ways it applies the foraging techniques at its disposal. Because there was no certainty that other members of kea society would necessarily know which foods were currently best to forage on or even how best to make use of them, keas were selected primarily for individual, as opposed to social, learning. Only by exploring and manipulating the objects in their environment could keas maintain the breadth of their diet and be poised to exploit any new food source that became available.

In essence keas were selected to play, since only through play could the requisite level of individual flexibility be achieved. In fact, all the most salient features of the kea's personality— its boldness, destructiveness, curiosity, and amusement—prove, when closely examined, to be aspects of play. It is as if keas had responded to the erratic alpine habitat by evolving a manic open-mindedness, a compulsion to play with anything in their surroundings, thereby ensuring the range of experience that would enable them to survive.[35]

SIX

From Bounties to Black Markets

> Conservation is not just a question of endangered species. New Zealand has many species that are still locally common but whose populations are small and isolated. In the face of continuing habitat loss and habitat modification, their future is uncertain. The endangered kakapo and kaka grab our attention and resources, while kea and parakeets may be sliding downhill to join them.
>
> Kerry-Jayne Wilson, "Kea: Creature of Curiosity"

SMUGGLERS' COVE

In 1989 the rangers at Arthur's Pass National Park asked us to record the license plate numbers of cars that showed up at the study site at odd hours of the day and any others whose drivers were not apparently disposing of trash. They had heard that keas, possibly poached from the park, were turning up for sale on the black market pet trade in Canada. A few months later, on

June 13, 1990, customs and conservation officials confiscated two suitcases at the Christchurch Airport that were bound for Singapore.[1] Inside were eight keas that had been drugged and then stuffed into plastic tubes. This was one of the few times that a smuggled shipment of native birds had been intercepted in New Zealand. By the time the suitcases were found, one of the birds was already dead. The mortality from this one smuggling effort was probably far higher, however. An estimated sixteen birds were initially stolen, with only eight surviving to be shipped. For every bird sold illegally on the international market, two to ten others will have perished, unable to endure the conditions along the way. Smuggled birds die from stress, suffocation, starvation, excess heat, excess cold, and overdoses of the drugs used to subdue them.

International bird smuggling is well organized and richly financed. This particular venture was planned by a smuggling ring that included Fredrick Robert Angell, a thirty-one-year-old New Zealander, together with a German national and several unknown contacts in Singapore. Wildlife officers in the New Zealand Department of Conservation had been notified that Angell, a convicted bird smuggler, had returned to New Zealand upon his release from prison in Australia. Hoping to forestall a new smuggling effort, department officers collaborated with the police and the United Nations agency that monitors the worldwide trade in endangered species to keep him under continual surveillance.

Angell could be convicted only with direct evidence that he was both stealing birds and attempting to smuggle them out of the country. By the time surveillance was established, Angell had already begun obtaining the keas. While under observation he broke into the Dunedin Zoo on the South Island and stole keas from their cages. At one point local police impounded Angell's car with a cage full of keas in the backseat. Wildlife officials convinced the police to return the car so that more evidence could be collected. In all, Angell took keas from at least three zoos and two national parks.

When the shipment was confiscated, Angell was sought out and arrested. He was eventually convicted on a variety of charges, but the legal penalties were relatively mild. For taking keas from Arthur's Pass and Fiordland National Parks, Angell admitted to two offenses of the National Parks Act, for which he received a total sentence of six months in jail. He also pleaded guilty to failing to provide reasonable accommodation for the birds, possessing and allowing a kea to go free, and possessing and attempting to export eight keas. These offenses carried no prescribed jail terms, however, and he was merely fined about $12,000 for his role in this smuggling attempt and a variety of earlier offenses.

After they were recovered, each of the seven surviving birds had to be somehow repatriated to an appropriate environment. The first challenge was to determine which animals had been raised in captivity, since most likely they would not be able to survive in the wild. Zoo officials eventually verified that two of the seven were captive birds, and they were soon returned to their cages. The five remaining keas were banded and released in Arthur's Pass National Park on June 25, 1990. One bird left immediately and was not subsequently sighted. Three others were observed intermittently over the following year. The fifth bird tried to join the group of keas that we had been watching at our study site.

When we arrived in December of 1990, we observed an adult

male with a strange banding pattern foraging at the site. Since we had banded our birds under a permit that included all other wild keas in New Zealand, we had a list of all the band combinations that were used by researchers. When we discovered that this bird's bands did not match any known scheme, we quickly checked with park officials to determine why. It was then that we were given the band combinations for the five new keas and told the story of the black market birds. The new bird, named "Metal-black" for his band colors, did not have an easy time of it. Kea social life is highly structured and dependent on a history of interactions with all other individuals in the group. Keas that have no history pay a harsh penalty.

This dark, fully mature male spent a large portion of his time at the study site. He clearly was not recognized as a member of the local group, and he lacked the social status usually accorded to adult males. Instead, he behaved like an unusually subordinate subadult, crouching in the presence of adult males, fleeing at their approach, and carrying what food he could obtain away into the woods to eat without disturbance. As a stranger he was attacked by every adult he approached and was displaced from food by even fairly subordinate individuals. In response he reverted to juvenile behavior and began hunching to other males, even juvenile males, in an effort to obtain access to resources. He was the only adult we had ever seen produce the hunch display. Although not highly successful at obtaining food in this fashion, he was eventually accorded some measure of tolerance by the resident birds. This may have been enough to help him survive.[2]

Reintroduction of animals from captivity poses serious problems, particularly in the case of highly social species like keas. This concern has long been recognized in captive breeding programs, where endangered species are reared in zoos and later released into the wild. But the difficulties can be almost as great in simply repatriating an animal that has been removed from a wild population. Surviving the social environment is often a key to foraging success, and strange individuals, or even familiar ones that have been temporarily absent from their accustomed group, lose status and are not allowed to forage at sites of rich food resources. Studying the social dynamics of a species can be critical to managing the health and safety of a social bird, particularly when individuals are to be introduced into an unfamiliar existing group.

KEAS IN THE COURTROOM

Smuggling for the pet trade has had a dramatic impact on many species of wild birds, particularly parrots. It rivals habitat destruction as a factor in the decline of parrot populations in South America, Africa, and Indonesia.[3] Export of many large parrots, including keas, is strictly forbidden, and few of these species breed well in captivity. As a result the demand for these exotic birds on the black market is so great that it effectively offsets the legal penalties that may be applied. Private collectors in Asia, North America, and Western Europe will reputedly pay up to $10,000 apiece for keas.

Until recently most wildlife managers believed there was practically no illegal trade in wild parrots from New Zealand. New Zealand law protects all native parrots from commercial trade, and the exports of this small country are carefully regulated. The smuggling incident in 1990, however, has brought attention to the vulnerability of keas, even where adequate laws and regulations are in place.

Compared to other nations, New Zealand has an exceptional history of protecting its native flora and fauna. The first legal protection for native animals was passed in 1907, safeguarding 36 species of native birds and the rare tuatara. The Animals and Game Protection Act of 1921 extended protection to native bats, frogs, and a further list of 132 native birds.[4] Legal protection for keas, however, came only relatively recently. They were the very last of the native New Zealand birds to be protected on both public and private land.

When other New Zealand species were being idolized by bird enthusiasts, keas were still being shot as pests. The *New Zealand Journal of Agriculture* records that 29,249 bounties for kea beaks were paid out between 1920 and 1929. The practice continued well into the 1940s. A report from the New Zealand Department of Agriculture indicates that bounties were paid on a total of 6,819 keas from 1943 through 1945. In 1953 the Wildlife Act was passed, giving keas limited protection, at least on public lands. On sheep ranches, however, they were still hunted with enthusiasm. By the best current estimate, the total number of keas killed between 1868 and the 1970s is 150,000.[5]

Toward the end of the 1960s, a voice of sympathy for the kea finally began to be heard among the New Zealand public. In 1970 the New Zealand government placed keas on the Fifth Schedule of the Wildlife Act of 1953, thus extending protection to keas on private lands. Landowners could now destroy the birds only if they were doing "injury or damage" to property. The law also prohibited capturing keas for sale as pets. In theory these changes represented a major advance in protecting the species. In practice, however, ranchers seldom had difficulty finding justification for killing keas, and enforcement of the act was spotty at best. About a hundred people were granted permits to keep keas in captivity, and some of these birds were used to attract other keas, which were then shot.

By the 1980s public demands mounted to make the kea a fully protected species. A major grass-roots campaign, spearheaded by the New Zealand Forest and Bird Society, led in 1986 to unqualified protection for keas, even on private lands. Today a host of national laws govern keas' safety within New Zealand borders.[6] In addition to the Wildlife Act of 1953, these include the National Parks Act, which forbids the removal of animals from parks and protected areas; the Animals Protection Act, which prevents mistreatment of wild animals; and the Trade in Endangered Species Act, which bans the export of any New Zealand parrot.

International trade in parrots is a closely regulated affair. New Zealand was the 103d signatory to the Convention in International Trade in Endangered Species of Wild Flora and Fauna (CITES) organized by the United Nations Environment Program. This treaty came into effect in 1975 to monitor the trade in many species of plants and animals, including all but two kinds of parrots (budgerigars and cockatiels). It prohibits the commercial trade of all endangered species of parrots across international borders and requires export permits for anyone handling species that are considered capable of tolerating commercial exploitation but may become threatened by trade.[7] Regulation, however, is but a first step toward protecting species from smugglers. Until the lucrative market for wild birds is curtailed, these laws can provide only limited protection.[8]

SHEEP AND KEAS TODAY

Wildlife officials believe that kea attacks on live sheep are uncommon today but still do occur. Nearly everyone we spoke to in the Department of Conservation knew of some high-mountain sheep run that reported damage to livestock from keas. These reports are not surprising, since the practice of leaving sheep flocks to forage on their own for long periods, without either a shepherd or a dog, is still common. In the mountains of the South Island, sheep are sent out to graze all winter and may not be seen for three or four months. Because of the kea's protected status, however, each report of damage by keas must now be investigated and documented by wildlife officers. The result is a far clearer and more balanced account of the frequency and severity of kea attacks on sheep. For example, in winter 1992 one ranch, Cora Lynn, lost six animals. Department investigators found that five of the dead sheep were quite old, and the sixth had been trapped in snow. These cases fit the general pattern of kea damage, which is mostly confined to animals that are old and sick and to the carcasses of sheep that died from other causes. Wildlife officials do, however, occasionally see healthy sheep with kea wounds.

Investigators from the Department of Conservation sometimes find it difficult to confirm the reports of ranchers. A runholder at one of the largest sheep ranches in the area called the department one spring to report that keas had killed three hundred of his sheep. When asked how he knew, he replied that he had just done a muster (a count), and three hundred of the animals were missing. There was, however, no physical evidence to substantiate his claim. In another example, in the fall of 1990, a rancher reported stock loss due to keas on his run. When wildlife officers visited the property the following day, the owner showed them one apparently healthy sheep with a small hole in its back in the kidney region. He stated that he and his musterers had found five other sheep killed by keas, each with a large hole in the loin area near the kidneys. He gave the specific location of each carcass, and the officers returned the next day, intending to find and examine the dead sheep and then trap the troublesome kea. At the first reported location they found no dead sheep, and none of the live sheep showed any evidence of kea attacks. At the second location they found one dead sheep, but it was too badly torn up by scavenging harriers to allow them to assess the cause of death. No keas were seen or heard during the investigation.[9]

Nevertheless, there is clear evidence that keas continue to inflict damage on sheep. At the Cora Lynn ranch, which borders on Arthur's Pass National Park, the average annual loss due to keas is estimated at about two sheep per thousand. The loss at other ranches may be somewhat higher, but the incidence of damage is quite irregular from one area to another and from one year to the next. Reliable estimates of the magnitude of the problem are difficult to obtain. Most kea attacks occur at night or in early morning, and the worst losses are in the winter, when the availability of natural kea foods is lowest and the sheep are often immobilized by snow. Kea damage is rarest after shearing, when the birds apparently cannot hold on to the sheep's backs.[10]

The nature of the damage can also be quite variable. The birds often remove tufts of wool, occasionally scratching or scraping the skin in the process. At the Flockhill run, wildlife officials have seen keas riding around on sheep's backs and pulling out their fleece. Little tufts of wool stand out on the backs of such

This kea was photographed feeding at night on a sheep carcass. (Photograph by Robin Smith/Department of Conservation, Christchurch, New Zealand)

animals and are visible at a distance. When keas do cause wounds, they are commonly not recognized until much later. Kea wounds on one young, healthy ewe were noticed when she was sheared, but by then they had healed on their own. Sometimes keas work a small hole in the hide by pulling and twisting. They then insert their upper bills and rotate them, scraping off the subcutaneous fat in a circle about 6–8 cm in diameter. They seldom penetrate the body cavity, but in rare cases live sheep have had wounds large enough to put a fist through. One sheep that died at Cora Lynn during the winter of 1990 showed a 5×8-cm hole in its back, located along the spine just forward of the kidneys.[11]

Other New Zealand birds also attack immobilized or dying

sheep or scavenge at sheep carcasses. Damage by these other species can be easily distinguished from kea-inflicted wounds. Black-backed gulls tend to work down through the back at the base of the sheep's rib cage. They cannot maintain a grip on healthy, mobile animals but will attack immobilized sheep immediately, usually pecking out the eyes first. Australasian harriers remove long strips of flesh from the neck and flank region of sheep carcasses. It is not known whether harriers ever attack living sheep.

Sheep that have been wounded by keas often die of blood poisoning caused by *Clostridium* bacteria. These tiny soil organisms are related to the bacteria that cause tetanus, botulism, and gas gangrene. There are two conflicting notions about how the disease works.[12] One theory suggests that *Clostridium* spores are widespread over pastureland and that sheep ingest them when grazing. The bacteria remain inside the sheep without causing harm. But if later the sheep are under physiological stress from poor nutrition or cold temperatures, any small scratch or wound can result in a major infection, symptoms of blood poisoning, and ultimately death. Keas, by pulling wool from live sheep and inflicting small wounds, might play a role in the series of events that activates the disease.

The second theory implicates keas more directly. It posits that because keas spend much of their time digging in the soil, they may come in contact with the bacterial spores, which would adhere to their bills. The kea would then inject the sheep with the disease when it pecked through the skin. Many high-mountain sheep ranchers favor this theory and have begun inoculating their sheep against blood poisoning.[13] Vaccination of sheep flocks minimizes or even eliminates mortality from the disease and thereby reduces winter mortality. Further studies are underway to determine the role of keas in triggering the disease.

There is a new and vocal generation of sheep farmers in New Zealand. Many of them appreciate seeing keas on their land, viewing these wily parrots as a proud symbol of the New Zealand high country. They accept the fact that some individual keas will occasionally cause damage to their stock and property, but they are willing to invest in protecting both their stock and the wild birds. In a recent popular article on keas, Philip Temple quotes an enlightened sheepman who commented philosophically on a 10 percent loss to keas in one flock: "Yes, it's been bad the last couple of years, but then some years there have been no losses to keas at all, most of the last fifteen, anyway. I'm prepared to take a few losses—it's one of those natural things." His attitude is heartening. It suggests that protection for the kea may ultimately help both stockmen and wildlife managers focus on practical solutions that do not require exterminating the birds.[14]

LEARNING TO LIVE TOGETHER

The current popular image of keas is generally positive. They are thought of as the Clowns of the Mountains: mischievous and amusing, bold and inventive. Their antics have become the stuff of modern folklore.[15] After all, what other bird has the audacity to shred windshield wipers under the noses of park service staff? Or dance on the roofs of tourists' cars? Or steal TV antennas from the houses of local villagers?

New Zealanders commonly refer to keas as "cheeky," which is surely an understatement. Not only are they bold, exploratory, and infuriatingly persistent, but they are amazingly destructive of human property. Today the kea's property crimes present as difficult a management challenge as its interactions with sheep. The birds evolved their foraging habits among rugged mountains and fragile alpine flora. But a conspicuous feature of their present-day ecology is their attraction to human habitations and their tendency to forage in sites where the major sources of food are of human origin. Keas are now more reliably found around garbage dumps, parking lots, ski lodges, and backcountry cabins than in any of the wilder and more remote locations of the Southern Alps.[16] Despite the certain dangers of associating with people, they continue to make human artifacts a focus of their activities.

All objects in national parks are subject to depredation by keas. At a campground at Arthur's Pass National Park, the birds shredded fourteen tents in the first two months of one summer season. They rifle neglected backpacks and demolish or steal their contents. The sturdiest of camping gear is no match for persistent keas. One evening we watched a group of juvenile keas attack a pair of heavy hiking boots, pulling out the laces and shredding the inner lining. By the next morning they had separated the thick lug soles from the leather uppers. Even relatively heavy objects are no obstacle, as attested in one description of a kea dragging away a $4\frac{1}{2}$-pound ax.[17] One kea we watched rolled up a large doormat and pushed it down a flight of stairs. Keas regularly pull over garbage cans and strew the contents across the countryside. They break off pieces of television antennas and put holes in rain gutters that make them look as if they had been attacked with a can opener. In kea country nothing that can possibly be removed or demolished is left intact.

Automobiles are a favored target. Keas of any age converge rapidly on camper vans and immediately begin to strip the rubber weather seals off the roof ventilators. They will approach any parked car and remove the wiper blades, windshield seals, ski racks, and radio antennas. They readily peel off vinyl trim and are even reputed to bite off the fill valves and let the air out of tires. In one horrendous case, a couple of backpackers left their soft-top jeep in a parking lot in Arthur's Pass. They returned five days later to find the roof ripped to shreds, the seat cushions demolished, and all the wiring torn out from under the dashboard. The car could not be started and had to be towed back to Christchurch.[18]

Tourists are often thoroughly entertained by these antics, particularly when the damaged goods belong to someone else. We once watched a crowd of perhaps a dozen tourists laughing and taking photos of a group of juvenile keas who were boldly wrecking an expensive ski rack on a parked car. When the owner of the vehicle returned, he was, needless to say, somewhat less amused.

Sometimes the birds' rampant destruction can create significant human hazards. Keas are a serious problem at ski areas. Not only do they make off with sunglasses and shred down parkas, but they also peel off the protective pads around the base of ski-lift towers, remove wiring for outdoor lighting and loudspeakers, and dismantle ski-lift warning systems. The Department of Conservation has taken measures to protect the ski areas along with the keas, with limited success. Although the officers captured and removed many keas that were plaguing the ski areas, the birds would often return, even after having been transported many kilometers away. In 1989 the department also established a

Keas are known for destroying automobiles, an amusing habit with serious consequences. (Illustration by Earl Tutty)

twenty-four-hour hotline for ski operators and others to report kea-related problems.[19]

The Park Service staff and the year-round residents of Arthur's Pass Village, a hardy group of about forty souls, have developed their own suite of kea defenses. Residents wire their garbage cans down and anchor them with concrete blocks, cover their television antennas with polyvinyl chloride piping, and close off chimneys and other attractive openings with chicken wire. Unattended backcountry sheds are often sheathed in steel siding, and the wiring and motors on ski lifts are protected with steel conduit and heavy metal casings. The head ranger at Arthur's Pass National Park drives a car that looks more like a submersible than a government vehicle, since the windshield wipers, radio antenna, and other "soft" parts are shielded with polyvinyl chloride.

In 1989 the department initiated a "Don't Feed the Kea" campaign. The central premise of this management strategy is that if people stop directly feeding keas, the birds will eventually cease

The head ranger at Arthur's Pass National Park has put PVC piping on the antenna and other vulnerable parts of his car to protect it from keas. (Photograph by Judy Diamond)

to associate people with food and will reduce their depredations on human property.[20] Residents have been more than willing to comply with the department's policy, and there is an active effort to educate tourists to refrain from luring keas into picnic areas and campsites.

Contact between keas and people is not just damaging to human property. Keas that frequent human settlements are often risking their own lives. Refuse dumps are particularly dangerous places, not just because of the toxicity of their contents. Keas can get entangled in string or discarded fishing line and break wings or legs as they attempt to free themselves. We have seen birds with monofilament fishing line wrapped so tightly around

In January 1989 the New Zealand Department of Conservation launched a campaign to discourage people from feeding keas. (Photograph by Judy Diamond)

one foot as to cause permanent crippling. Several birds we captured showed signs of swelling in the feet, a condition called "bumble-foot" in caged birds. The swelling results from infections, probably due to cuts from broken glass and metal fragments in the dump. We have captured adults with eroded areas at the base of the upper bill, apparently the result of trauma and subsequent infection. We saw several individuals that had lost a leg, perhaps through tangling or infections, or possibly through being trapped between the sharp-edged lid and the side of a discarded tin can. Near the study site, over a two-year period, park rangers found the bodies of twelve keas, some struck by passing cars on the nearby highway and one hit by a train.[21]

Such hazards have led the Department of Conservation to place high priority on finding an alternative, secure method of garbage disposal or locating a suitable site outside the park for the refuse dump.[22] During our research study, the dump at Arthur's Pass was the only unregulated open dump in the high country of the South Island. It was finally permanently closed in 1997. Although its closure reduces the birds' access to many dangerous materials, it also presents new conflicts. Keas are known to have fed at the dump for over forty years, with a local population of about seventy birds making use of it during the summertime. Generations of keas have grown to see the dump as a rich and reliable food resource. Wildlife managers thus anticipate that at first more keas will stray into the village to search for easy substitutes, and harassment of visitors and destruction of property may temporarily increase. In the long run the absence of the dump may cause the local kea population to restabilize at a lower level.

Keas may also be affected by attempts to eradicate the introduced animals that are damaging native habitats throughout New Zealand. Australian possums are still very common in New Zealand, even in the face of drastic control measures.[23] Since the 1950s managers have attempted to poison them by lacing bait with 1080, a poison made from a nerve gas developed in World War II. After the war, 1080 was patented in the United States as a rodenticide and was widely used by ranchers throughout the American West to kill coyotes. Completely odorless and tasteless, it is an effective killing agent. But the poison remains in the tissues of dead animals, which then serve as toxic baits to other animal species. In 1972 the United States prohibited the use of 1080 on public lands, in part because of evidence that it was killing large numbers of golden eagles and red-tailed hawks.[24]

Unfortunately, 1080 has been used in areas where keas occur, and the birds have been seen feeding on 1080-laced carrots on the ground. Their interest in carrion probably exposes them to additional poisoning through tainted possum carcasses. The risks involved in the use of 1080 in kea habitat are considerable, since the population is not likely to recover quickly if reduced by poisoning.[25] Efforts are now being made to restrict the use of the poison, but no other control method, including saturation trapping and hunting with specially trained dogs, has proved wholly effective in eliminating possums on the South Island.

HOW MANY KEAS REMAIN?

In 1990 we talked to an old-timer who had recently hiked through a section of remote backcountry in Arthur's Pass National Park. The route was one he had traveled many times before. In the 1960s he would generally see several keas during each day's travel on this trail. This time, in the course of a week of walking, he saw only one. The following year, on the same trek, he saw no keas at all. Are keas less common now, even in the national parks? Will the species soon become endangered? The issue is crucial to conservation and management of the species.

The populations of all New Zealand parrot species have declined to at least some degree over the past century. The kakapo is on the brink of extinction, with only a handful of the birds remaining on offshore island sanctuaries. Kakas are rare on the North Island, and their numbers are also declining on much of the South Island. Red-crowned parakeets are scarce on the mainland, and some island subspecies are extinct; the yellow-crowned parakeet is now much restricted in distribution; and the orange-fronted parakeet is known only from mid- and northwest Nelson. A close cousin of the kea, the Norfolk Island kaka provides a bitter reminder of the fragility of parrot populations. This bird was originally found only on a few isolated islands scattered between New Zealand and New Caledonia, but the last individual died in captivity in 1851. The introduction of rats and nonindigenous plants and birds is thought to have caused the species' decline; the birds were also shot by settlers for food. The Norfolk Island kaka is one of only seven species of parrots that have been driven to extinction in modern times.[26]

Today the kea is the only native parrot that can be readily observed at many localities on the mainland of New Zealand. This visibility makes it seem that the species is fairly abundant and that the keas that visit tourist locations are an overflow from a much larger population in the hinterlands. In fact the total population of any species, particularly one spread thinly over a great expanse of trackless wilderness, is extraordinarily difficult to estimate. Keas, moreover, present a special problem in this regard. Many of the techniques that biologists use to estimate population assume that the species is evenly distributed throughout its range, so that the results of a local bird count can be generalized to other, more remote regions with the same type of habitat. But since keas are attracted to people, even if there are only a few of the birds in a given area, they may readily congregate around observers. Their distribution is therefore likely to be clumped and irregular, with large aggregations in areas that are frequented by people.

There has been a series of attempts to estimate the overall kea population. Kerry-Jayne Wilson has banded keas in several national parks, particularly Mount Cook. She believes that about 60 percent of the birds around Mount Cook regularly beg for food in parking lots. By her estimate there may be no more than two hundred keas in the Mount Cook area and perhaps only two thousand in all of New Zealand, few of them outside protected areas. An earlier estimate, by the New Zealand Wildlife Service (now the Department of Conservation) in 1986, was roughly five thousand keas in all.[27]

To improve the precision of such estimates, it is necessary to take the distribution of the species into account. The primary source of information about bird distributions in New Zealand is a database of sightings developed by the Ornithological Society of New Zealand.[28] Between 1969 and 1985, members of the society made a note of all bird species seen on each of their field trips and in which map quadrat they were seen, with the quadrats being indicated by the 10,000-yard squares of the New Zealand National Grid. We recently analyzed these data to determine the probability of sighting keas during a single visit to each of the quadrats of the South Island; from that we estimated the wild population to be about 3,000.[29]

The probability of sighting keas is displayed on the map on page 144. It expresses the distribution of kea population density across the South Island, revealing several intriguing patterns. The areas in which one is almost certain to encounter a kea in the course of a single visit are scattered along the Southern Alps. There is a very small area of high density in the backcountry of Nelson Lakes National Park, a patch of high density running from Arthur's Pass National Park and Craigieburn Forest Park south and west along the mountain ranges, and a localized kea hot spot at Mount Cook National Park. The primary center of kea abundance, however, is in the southwest, in an area running from Westland National Park through Mount Aspiring

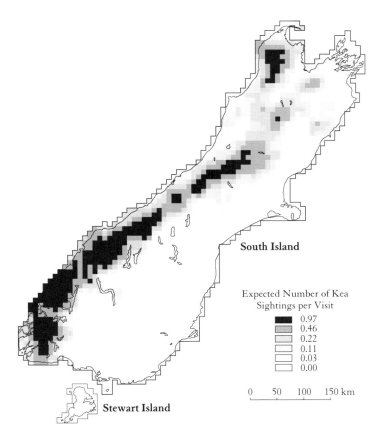

The distribution of keas in New Zealand. The darkened areas indicate regions in which keas were sighted at some time during a twenty-year survey of the bird fauna of the South Island by the Ornithological Society of New Zealand. The two darkest shades delineate areas in which a traveler might see keas at least once in every three visits. Keas are most densely concentrated in a band along the Southern Alps, primarily in national and regional parks; compare with the map on page 29. (Map by Alan Bond)

National Park to Fiordland National Park. A secondary region of high kea density is in the Tasman Mountains in Kahurangi National Park, in the northwestern corner of the island.

Note that keas are uncommon to rare outside government preserves and national parks—the result, in part, of a century of persecution by bounty hunters, as well as the modification of high-country habitat to make room for sheep. Keas were common in the mountainous regions of Otago and south Canterbury in the southern and central portions of the island in the late 1800s, but there are virtually none there now. Note also that the distribution of keas roughly corresponds to that of southern beech forests in the South Island, which confirms the reportedly heavy reliance of keas on the beech habitat.[30]

Kea populations have apparently declined since the nineteenth century. When Marriner wrote the first monograph on the kea in 1908, the birds were already much less abundant than in the 1800s. He commented that "in some districts, where they were once to be seen in large flocks, the long slaughter has since greatly reduced their numbers." And the decline appears to have continued well into the mid-twentieth century. Bounty hunters in the 1920s and 1940s killed more keas than are estimated to live today. But the decrease in the kea population was not wholly the result of persecution by sheep ranchers. Changes in husbandry practices, such as destroying carcasses and ceasing to dispose of offal in open pits, may have impacted kea populations almost as much as the bounty-hunting campaign.[31]

It is not clear whether kea populations continued to decrease even after the species was given some measure of protection in 1953. One 1976 survey suggests that populations in protected areas are at least holding their own. More than thirty runholders with lands adjacent to the Routeburn Basin in Fiordland National Park reported no change in the size of kea flocks over recent years. In addition, by marking and recapturing keas in Arthur's Pass National Park in the early 1990s, we obtained estimates of population density consistent with those that Jackson had recorded for the same area thirty years earlier.[32]

Some observers have noted an apparent decrease in numbers since the 1960s, but it may owe to a historical anomaly. During the "Deer Wars" of the 1960s, when the Department of Conservation was eliminating the herds of red deer that had laid waste to the New Zealand forests, deer carcasses were liberally scattered across the backcountry. Keas frequently fed on deer carrion, and the supplemental protein may have produced a temporary boom in the local population. This may be why more keas were observed in the 1960s than at present.

New Zealand wildlife officials are seeking to develop a management plan for the kea that relies on an accurate estimate of the population. With this issue in mind, the IUCN/SSC Captive Breeding Specialist Group conducted a population viability analysis of both keas and kakas in 1991. The final report of this meeting, including simulations of the population dynamics of both species, suggested that keas were not currently endangered but that kakas might well be, at least on the South Island.[33] There are plans for a national kea census, which would provide an additional source of reliable numbers.

Throughout the kea's history, its populations have fluctuated widely. Dramatic increases and decreases coincided with other major geographic and ecological events. Evidence suggests that keas flourished between fourteen thousand and ten thousand years ago, when the glaciers were slowly retreating and the shrub lands were expanding on the South Island. Then, about nine thousand years ago, the kea population started to shrink, possibly in response to the reinvasion of competitors, including its forest-dwelling cousin, the kaka. A thousand years ago, when the first humans arrived in New Zealand, keas may have been relatively uncommon, since the early settlers made little note of them. Not until sheep ranching provided an abundant new food source did kea populations expand again. Since 1900, with more careful sheep management practices resulting in reduced food, the population has once again reached a lower level. In the national parks, at least, this level appears stable.

THE EVOLUTION OF FLEXIBILITY

The kea originally came to the attention of nineteenth-century scientists as a conundrum, a vegetarian species that rapidly converted to carnivorous habits. To the scientists of that time, it was remarkable that an organism's features could evolve for one set of purposes and then be adapted completely to another. But we reinterpret the kea's evolution to suggest that its apparent conversion is only one aspect of the innate versatility of the species.

Keas originally foraged on the native plants and animals of the alpine environment and scavenged moa carrion when it was available. When sheep were introduced to the New Zealand high country, keas were probably initially attracted to them as objects of play. That they could grab on to the fleece on the sheep's backs and ride around on them must have made these woolly toys irresistibly entertaining. Keas probably passed much of their time around sheep, playing with them as with all the objects associated with sheep ranching. There would have been many opportunities for adult keas to feed on sheep carcasses, on dying or disabled sheep, and on sheep immobilized in snow. Having played with sheep as fledglings and juveniles might have made these adults more likely to approach sheep once they had gained expertise in foraging. Viewing sheep as a food resource in addition to a source of amusement, keas would have maintained their interest in sheep and perpetuated it through social facilitation.[34]

Thus the kea's two innate predispositions, the compulsion to play and the tendency to socially facilitate, orchestrated its interactions with sheep in the same fashion that they have helped to shape all aspects of the kea's natural history. In combination they constitute a mechanism for the generation of flexibility, for rapid adaptation to novel circumstances.

Ernst Mayr, the dean of American evolutionary biologists, has referred to species that emphasize behavioral flexibility as "open-program" animals, which typically lack predetermined patterns of behavior.[35] The distinction has less to do with the ability to learn than with the animal's orientation toward learning; such individuals actively seek out opportunities for acquiring new skills and making use of novel materials. Their motivation is not necessarily a desire for new information, however. The behaviors associated with flexibility may in themselves be the primary motivation. In other words, an animal plays because play is its own reward, not because play will result in learning. Similarly, an animal attends to what a social partner is doing not so much to learn from the other's actions as to obtain a share of a resource.

Open-program animals evolve in response to a particular constellation of environmental features. Not only is their environment unpredictable, but it continues to be so throughout the life of the animal. Other animals born into environments that are different from those of their parents must also be prepared to adapt to new conditions, but once having adapted, they do not need to keep altering their behavior. The kea, however, does. Its environment presents a set of continually changing circumstances—a situation in which play should evolve in its most striking form.[36]

Social facilitation helps young keas focus their attention on the foraging success of other individuals, a habit that ultimately provides a safety net for the young, ensuring their survival. They rely on older keas to indicate what food is available, but at the same time new resources are continually arising, and existing ones are frequently being depleted. Thus, young keas must divide their attention between the activities of older animals and their own playful exploration of new resources.

These conditions do not favor the evolution of a system of cultural transmission. Other parrots have been shown to exhibit observational learning and imitation, at least in captivity, which suggests that keas ought to be similarly capable of learning skilled behavior from other birds. But their foraging actions are very simple; young keas readily master them without having to learn from their elders. Foraging skills are honed primarily through trial and error, and observational learning plays a minimal role.[37]

One can create a sort of evolutionary "recipe" to describe the factors that promote flexibility in animals. For the kea, the first ingredient is a lineage predisposed to sociality. Next comes the relatively low food abundance, so the animal has to be prepared to accept an enormous range of foods. Add to this the short-term limitations on food availability, so the animal must constantly shift food sources. And finally, include delayed maturation and lenience by adults toward the young, so the young have time to play.[38] Against the dramatic background of the formation of the New Zealand islands, this recipe transformed a rather ordinary forest parrot into one of the world's most unusual birds, a species so flexible that it managed to survive even the extermination of much of its original ecosystem.

Flexibility, however, clearly has its limits. Although keas appear able to consume virtually any foodstuff that normally occurs in their habitat, as well as a great deal of exotic material introduced by people, they have never shown an inclination or ability to move beyond their mountain retreats. Their core allegiance to the high-mountain beech forest may be their only truly inflexible feature. Even today, keas feed extensively on beech fruit and foliage at all seasons, they seldom travel far from beech forest, and they almost never breed anywhere else. Although they could probably make a living in the mountainous areas of the North Island or even among the introduced pines and eucalyptus on tree plantations, keas are never found there. Whether because of the relentless hunting pressures they encountered on private lands or because of an unbreakable linkage to mountain beech in at least some aspects of their lives, keas have been unable to move out from their alpine refuges. One suspects that even open programs have their limits, and a strategy that provides great flexibility in one particular habitat may not be as effective in others.

Humans are the most extreme example of an open-program species. As we are now in the process of completely transforming our own habitats, it may be wise to consider the kea's limitations and ponder whether we, too, may have limits to our adaptability, limits that will shape the future ecology of this planet.

APPENDIX A

Common and Scientific Names of the Animals and Plants Mentioned in the Text

BIRDS

NATIVE TO NEW ZEALAND

adzebill	*Aptornis otidiformis*
bellbird	*Anthornis melanura*
cuckoo, long-tailed	*Eudynamys taitensis*
eagle, Haast's	*Harpagornis moorei*
falcon, New Zealand	*Falco novaeseelandiae*
fantail	*Rhipidura fuliginosa*
goshawk, New Zealand	*Circus eylesi*
gull, black-backed	*Larus dominicanus*
harrier, Australasian	*Circus approximans*
huia	*Heteralocha acutirostris*
kaka	*Nestor meridionalis*
kakapo	*Strigops habroptilus*
kea	*Nestor notabilis*
kiwi, little spotted	*Apteryx owenii*

kokako	*Callaeas cinerea*
moa	*Anomalopteryx didiformis*
	Dinornis giganteus
	D. novaezealandiae
	D. struthoides
	Emeus crassus
	Euryapteryx geranoides
	E. curtus
	Megalapteryx didinus
	Pachyornis australis
	P. elephantopus
	P. mappini
morepork	*Ninox novaeseelandiae*
New Zealand wrens	
bush wren	*Xenicus longipes*
rifleman	*Acanthisitta chloris*
rock wren	*Xenicus gilviventris*
owl, laughing	*Sceloglaux albifacies*
owlet-nightjar	*Aegotheles novaezealandiae*
parakeets	
red-crowned	*Cyanoramphus novaezelandiae*
yellow-crowned	*C. auriceps*
pigeon, New Zealand	*Hemiphaga novaeseelandiae*
piopio	*Turnagra capensis*
pipit, New Zealand	*Anthus novaeseelandiae*
pukeko	*Porphyrio porphyrio*
raven, New Zealand	*Corvus moriorum*
robin, New Zealand	*Petroica australis*
saddleback, North Island	*Philesturnus carunculatus rufusater*
silvereye	*Zosterops lateralis*
snipe-rail	*Capellirallus karamu*
stitchbird	*Notiomystis cincta*
takahe	*Porphyrio mantelli*
tui	*Prosthemadera novaeseelandiae*
weka	*Gallirallus australis*
whitehead	*Mohoua albicilla*
yellowhead	*M. ochrocephala*

INTRODUCED TO NEW ZEALAND

blackbird	*Turdus merula*
chaffinch	*Fringilla coelebs*
guinea fowl, tufted	*Numida meleagris*
sparrow, house	*Passer domesticus*

OTHER

blue jay	*Cyanocitta cristata*
budgerigar	*Melopsittacus undulatus*
cardinal, northern	*Cardinalis cardinalis*
chickadee, black-capped	*Parus atricapillus*
cockatiel	*Nymphicus hollandicus*
condor, Andean	*Vultur gryphus*
conure, Carolina	*Conuropsis carolinensis*
corella, long-billed	*Cacatua tenuirostris*
crow, American	*Corvus brachyrhynchos*
eagle, golden	*Aquila chrysaetos*
emu	*Dromaius novaehollandiae*
falcon, peregrine	*Falco peregrinus*
grebe, great crested	*Podiceps cristatus*
gulls	
herring	*Larus argentatus*
kelp	*L. dominicanus*
laughing	*L. atricilla*
ring-billed	*L. delawarensis*
hawk, red-tailed	*Buteo jamaicensis*
kaka, Norfolk Island	*Nestor productus*
macaws	
Cuban	*Ara tricolor*
glaucous	*Anodorhynchus glaucus*
magpie, black-billed	*Pica pica*
nighthawk, common	*Chordeiles minor*
oriole, northern	*Icterus galbula*
ostrich	*Struthio camelus*
parakeets	
black-fronted	*Cyanoramphus zealandicus*
Newton's	*Psittacula exsul*
Seychelles	*P. wardi*
raven, common	*Corvus corax*
turkey	*Meleagris gallopavo*
whippoorwill	*Caprimulgus vociferus*

REPTILES and AMPHIBIANS

NATIVE TO NEW ZEALAND

frog	*Leiopelma* sp.
	Litoria sp.
gecko	*Heteropholis* sp.
	Naultinus sp.

skink   *Leiolopisma* sp.
tuatara   *Sphenodon punctatus*

MAMMALS

NATIVE TO NEW ZEALAND

bat, New Zealand short-tailed   *Mystacina tuberculata*

INTRODUCED TO NEW ZEALAND

cat, domestic   *Felis catus*
deer, red   *Cervus elaphus*
dog, domestic   *Canis familiaris*
goat, domestic   *Capra hircus*
mouse, house   *Mus musculus*
pig, domestic   *Sus scrofa*
possum, brush-tailed   *Trichosurus vulpecula*
rabbit, European   *Oryctolagus cuniculus*
rats
  black   *Rattus rattus*
  kiore   *R. exulans*
  Norway   *R. norvegicus*
sheep, domestic   *Ovis aries*
stoat   *Mustela erminea*

OTHER

bison   *Bison bison*
capuchin monkey, tufted   *Cebus apella*
coyote   *Canis latrans*
elk, Canadian   *Cervus canadensis*
macaque, Japanese   *Macaca fuscata*
sheep, bighorn   *Ovis canadensis*

FISH

NATIVE TO NEW ZEALAND

barracouta   *Thyrsites atun*

INVERTEBRATES

NATIVE TO NEW ZEALAND

blackfly, New Zealand   *Austrosimulium australense*
peripatus   *Peripatoides novaezealandiae*

scale insect	*Ultracoelostoma assimile*
snail, land	*Powelliphanta* sp.
wetas	
bush	*Hemideina* sp.
giant	*Deinacrida* sp.

PLANTS

Native to New Zealand

astelia	*Astelia linearis*
broom, native	*Carmichaelia* sp.
coprosma	*Coprosma* sp.
daisy, mountain	*Celmisia* sp.
flax, mountain	*Phormium cookianum*
grass, tussock	*Chionochloa* sp.
hebe, willow leaved	*Hebe salicifolia*
kauri	*Agathis australis*
lily, mountain	*Ranunculus lyallii*
rata, southern	*Metrosideros umbellata*
rimu	*Dacrydium cupressinum*
snow groundsel	*Senecio scorzoneroides*
southern beeches	*Nothofagus* sp.
beech, mountain	*N. solandri* var. *cliffortioides*
beech, silver	*N. menziesii*
spear-grass	*Aciphylla* sp.
totara, mountain	*Podocarpus nivalis*
tree daisy	*Olearia quicennifolia*
tree tutu	*Coriaria arborea*
vegetable sheep	*Haastia pulvinaris*
	Raoulia eximia
	R. mammillaris
	R. rubra

Introduced to New Zealand

eucalyptus	*Eucalyptus* sp.
pine, Monterey	*Pinus radiata*

REFERENCES: Falla et al. 1978; Feduccia 1996, 323; Heather and Robertson 1997; King 1984, 217–20; McMillan, Hart, Robertson, Stewart, and Webber 1989; O'Brien 1981, 155–57; Salmon 1985; Stewart 1984.

APPENDIX B

Supplementary Tables

TABLE B1. Kea Population at the Study Site, by Age Class

Age Class	1989	1990	1991	Mean	Standard Deviation	Coefficient of Variation (%)
Adult	38.1	35.9	35.2	36.4	1.51	4.2
Subadult	7.1	8.9	14.5	10.2	3.86	37.8
Juvenile	18.4	8.0	34.2	20.2	13.19	65.3
Fledgling	—*	7.3	6.5	6.9	0.32	4.6

NOTE: These estimates are based on resightings of banded individuals during hourly censuses, derived from table 1 in Bond and Diamond 1992, 153.

*Censuses were conducted prior to the emergence of fledglings in 1989.

TABLE B2. Foraging Activity, by Sex and Age Class

Sex	Age Class	Sample Size	Excavating	Demolishing	Scraping	Searching	Eating	Transporting	Stealing	Gleaning
Male	Adult	22	0.09561	0.05792	0.04353	0.17543	0.37957	0.06535	0.01437	0.12816
	Subadult	8	0.06019	0.11658	0.09616	0.18468	0.29104	0.07467	0.02611	0.16520
	Juvenile	14	0.03925	0.11559	0.07703	0.16140	0.27246	0.04434	0.00779	0.14034
	Fledgling	5	0.02143	0.21683	0.16584	0.19096	0.07019	0.03449	0.00142	0.26165
Female	Adult	2	0.03267	0.08659	0.06896	0.15898	0.06917	0.09003	0.02439	0.23956
	Subadult	3	0.08969	0.16354	0.14584	0.18727	0.13287	0.08849	0.01649	0.25353
	Juvenile	4	0.01464	0.07303	0.03868	0.22043	0.21436	0.04615	0.00362	0.11735
	Fledgling	2	0.02349	0.25390	0.23350	0.11472	0.09417	0.07540	0.00896	0.21922

NOTE: We classified the birds' behavior into "activity" categories based on the occurrence of component action patterns. For each thirty-second interval in the data, we created a binary record, a sequence of ones and zeros, that indicated the presence or absence of each action pattern. To characterize the activity categories, we generated a set of weighting coefficients that indicated the relative importance of each behavior to the activity in question. The coefficients were derived in part from a cluster analysis of the binary records (Bond and Diamond 1995). Each record was then multiplied by the weighting coefficients characteristic of each activity in turn, generating one score for each activity for each binary record. We averaged the scores for each bird and considered the result as the level of performance of that activity for the given individual. For each activity, the table entry is the mean performance across all individuals of the same sex and age class. The sample size is the number of individuals in the given sex and age class. The significance of differences between classes was determined with a two-way analysis of variance (sex × age) on each activity grouping, with subsequent specific contrast tests.

The analysis reveals the following results:

(continued)

TABLE B2 (*continued*)

Excavating. Adult males spent significantly more time than juvenile and fledgling males; among females, subadults spent significantly more time than any other age group.

Demolishing. The time spent differs significantly by age but not by sex. All older age groups, male and female, spent significantly less time in this activity than fledglings. In addition, subadult females spent more time than juvenile females, and adult males spent less than either subadult or juvenile males.

Scraping. Adult and subadult females spent less time than fledglings, while juvenile females spent less time than subadult females. Among males, all three older age groups spent significantly less time in this activity than fledglings; furthermore, adults spent less time than subadults.

Searching. Female juveniles spent more time than male juveniles and female fledglings. All other differences were not statistically significant.

Eating. Among adults and subadults, males spent more time than females. Among males, the time spent varied greatly by age, with adults spending significantly more time than subadults and juveniles, which in turn spent significantly more time than fledglings. Age differences among females were not statistically significant.

Transporting. The time spent did not differ significantly across age class or sex.

Stealing. Although the incidence was highly variable across individuals, theft was significantly more frequent among subadults and adults, pooled across sex, than among juveniles and fledglings. Among males, subadults and juveniles stole more often than fledglings.

Gleaning. Among males, all older age categories spent less time in this activity than fledglings. The differences among age groups in females were not significant. Among adults and subadults, females spent significantly more time than males.

TABLE B3. Proportion of Time Spent at the Study Site, by Age Class and Sex and by Age Class and Time of Day

	By Sex			By Time of Day (Both Sexes)				
	Male	Female	Both	5–8 A.M.	8–11 A.M.	2–5 P.M.	5–8 P.M.	8 P.M.–12 A.M.
Adult	0.108	0.039	0.105	0.107	0.061	0.095	0.151	0.111
Subadult	0.160	0.115	0.152	0.183	0.015	0.090	0.211	0.263
Juvenile	0.232	0.081	0.209	0.220	0.103	0.143	0.276	0.301
Fledgling	0.139	0.306	0.181	0.268	0.113	0.093	0.187	0.117
All Ages	0.160	0.093	—	0.170	0.068	0.107	0.213	0.212

NOTE: From hourly censuses of banded birds at the site, we estimated the proportion of time each individual spent in the area based on the proportion of censuses in which it appeared. The table displays mean proportions within category groupings; significance of differences was assessed with analysis of variance. Sample sizes were as follows: adults, 60 males and 3 females; subadults, 10 males and 2 females; juveniles, 16 males and 3 females; and fledglings, 6 males and 2 females.

The analysis reveals the following results:

Sex Differences. Males and females do not generally differ in the proportion of time they spend at the site, except among juveniles. Juvenile females spend significantly less time in the area than juvenile males.

Age Class Differences. Juvenile and fledgling keas spend a significantly larger proportion of their time at the study site than adults do.

Diurnal Pattern. Keas spend significantly more of their time at the site in the early morning, between 5:00 and 8:00 A.M., and in the evening, between 5:00 P.M. and midnight, than they do at any other time of day. They rest through much of the early afternoon (from noon to 2:00) and are seldom seen at the study site during this period.

Age Class Differences in Diurnal Pattern. In the morning session, fledglings and juveniles spend more time at the site than adults, with subadults in an intermediate position. Later in the day the differences between age groups decline, along with the total proportion of time spent. After 5:00 P.M. the investment of time picks up again, with juveniles spending more time than adults; subadults also spend increased amounts of time during the evening session. The late evening hours, from 8:00 to midnight, are dominated by juveniles and subadults, with both adults and fledglings spending significantly less time in the area.

TABLE B4. Social Activity, by Sex and Age Class

Sex	Age Class	Sample Size	Hunching	Attacking	Threatening	Crouching	Toss Play	Tussle Play	Locking Bills	Play Biting
Male	Adult	22	0.00152	0.02903	0.02922	0.00194	0.00230	0.00052	0.00000	0.01625
	Subadult	8	0.00140	0.01675	0.01944	0.02054	0.00084	0.00105	0.00000	0.00836
	Juvenile	14	0.03264	0.04359	0.03310	0.01621	0.00149	0.01052	0.00301	0.01878
	Fledgling	5	0.02880	0.06736	0.02184	0.02049	0.00216	0.00958	0.01078	0.01690
Female	Adult	2	0.00000	0.03392	0.01131	0.02439	0.00962	0.00842	0.00581	0.00551
	Subadult	3	0.00000	0.04720	0.00000	0.02282	0.01426	0.01718	0.01449	0.01262
	Juvenile	4	0.00807	0.02613	0.01386	0.00541	0.00106	0.00266	0.01064	0.00798
	Fledgling	2	0.00000	0.06452	0.01613	0.01042	0.00000	0.00870	0.04720	0.02895

NOTE: We classified the birds' behavior into "activity" categories based on the occurrence of component action patterns. For each thirty-second interval in the data, we created a binary record, a sequence of ones and zeros, that indicated the presence or absence of each action pattern. To characterize the activity categories, we generated a set of weighting coefficients that indicated the relative importance of each behavior to the activity in question. The coefficients were derived in part from a cluster analysis of the binary records (Bond and Diamond 1995). Each record was then multiplied by the weighting coefficients characteristic of each activity in turn, generating one score for each activity for each binary record. We averaged the scores for each bird, and considered the result as the level of performance of that activity for the given individual. For each activity, the table entry is the mean performance across all individuals of the same sex and age class. The sample size is the number of individuals in the given sex and age class. The significance of differences between classes was determined with a two-way analysis of variance (sex×age) on each activity grouping, with subsequent specific contrast tests.

The analysis reveals the following results:

Hunching occurred significantly less often among adult and subadult males than among fledglings and juveniles. The only females observed hunching were juveniles. The lack of significant difference in the incidence of this behavior between male and female juveniles may be due to the low sample size and high individual variability of hunch behavior among females. Only two of the four female juveniles we recorded showed any hunching behavior at all, whereas twelve of the fourteen male juveniles displayed the behavior, nine of them at higher levels than we observed for any of the female juveniles.

Attacking. Adult and subadult males attacked other keas significantly less often than fledglings, and subadult males attacked significantly less than juveniles. The differences in this behavior between fledgling and older females were not significant, presumably because there were few females in the sample, but these differences were in the same direction and of similar magnitude to the differences among males.

Threatening. Females appear to threaten their opponents less often than males, but this difference failed to achieve statistical significance.

Crouching. There was a significant difference in the incidence of this behavior between male and female adults and between adult and subadult males. Although older females appear to crouch more often than younger ones, the behavior was too rare and variable in females to allow meaningful statistical assessment.

Toss play is mainly an activity of older females. Adult and subadult females, pooled together, engaged in toss play more often than younger females. Among adult and subadult birds, females engaged in toss play more often than males.

Tussle play among males is mainly the province of younger birds. Adult and subadult males engaged in tussle play significantly less often than fledgling or juvenile males. Females engaged in tussle play both as fledglings and as subadults and adults.

Locking bills is significantly more frequent overall in females than in males. Among males, fledglings lock bills more often than adults or subadults.

Play biting, in contrast to locking bills, showed no significant difference by age or sex.

161

TABLE B5. Dominance Status as a Function of Age and Sex

	Interactions with Adults and Subadults			Interactions with Juveniles and Fledglings		
	Male	Female	Both	Male	Female	Both
Adult	0.053	−0.450	0.021	0.008	−0.382	−0.022
Subadult	−0.496	−0.857	−0.604	−0.016	−0.494	−0.136
Juvenile	−0.012	−0.726	−0.171	0.086	−0.535	−0.052
Fledgling	−0.310	−0.531	−0.374	−0.008	−0.223	−0.070
All Ages	−0.064	−0.676	—	0.026	−0.439	—

NOTE: We determined the dominance status of marked birds from the outcomes of social interactions during our data recording sessions as a "dominance index." The index is the number of interactions in which a given bird displaced another minus the number in which it was displaced, divided by the total number of its interactions. The index is positive for birds that tend to displace others and negative for those that tend to be displaced. To look for evidence of nonlinearity in the dominance hierarchy, we computed the index separately for interactions of each individual with adults and subadults, and then for interactions with juveniles and fledglings. The effects of sex and age group on dominance status were then assessed using analysis of variance. Sample sizes were as follows: adults, 31 males and 2 females; subadults, 8 males and 3 females; juveniles, 14 males and 4 females; and fledglings, 5 males and 2 females.

To test the legitimacy of our aggregate measures of dominance, we subsequently compared the observations of wins and losses between known individuals with what would have been predicted from their relative dominance indices. We observed 177 interactions between known individuals for whom there was enough additional data to estimate dominance indices. The relative dominance indices of the participants correctly predicted 75 percent of the outcomes.

The analysis reveals the following results:

Sex Differences. The dominance status of females was significantly lower than that of males, primarily because of differences between the sexes among juveniles and subadults. Adult and fledgling males did not display a significantly higher average dominance status than females of their same age, whether in interactions with adults and subadults or in interactions with juveniles and fledglings.

Age Differences. When interacting with adults and subadults, adults (males and females pooled) displayed a significantly higher dominance status than fledglings and subadults. Juveniles were also superior to subadults by this measure.

(*continued*)

TABLE B5 *(continued)*

Age Differences within Sex Categories. Examined alone, males showed the same pattern of age differences as was seen in the pooled data for interactions with adults and subadults. Females also exhibited a strong age dependency in mean dominance level. Adult females are significantly more dominant than juvenile and subadult females, and female fledglings are also more dominant than subadult females.

In contrast, interactions with juveniles and fledglings showed no significant age effects on dominance. If kea society contained no circular dominance relationships, the same pattern of age-related differences would be apparent no matter which interactions were used as a basis for assessing dominance. The absence of age differences in interactions with juveniles and fledglings thus confirms the lack of transitivity in the dominance structure.

TABLE B6.   Principal Play Behaviors in Keas

LOCKING BILLS

One kea grasps the other's culmen in its bill and twists and pushes, using its own body weight for leverage (Keller 1975). Although the behavior superficially resembles regurgitant feeding, no food is transferred, and the grasping bird does not shake or vibrate its bill.

BITING

The bill surrounds an object and holds it more or less tightly (Keller 1976). Keller observed a form of this behavior exhibited during courtship and play. We observed keas in play repeatedly grabbing a part of another individual—particularly the tail, feet, or legs—with their bills.

PUSHING WITH THE FEET

K. J. Potts (1969) includes in his definition of "clawing" the high-intensity behavior of raising a foot and striking out sideways, as well as a variety of lower-intensity variations. We commonly observed keas in real and play combat using a foot to push at another kea, which constitutes a low-intensity display of this behavior. High-intensity displays of clawing or kicking were common in aggressive attacks that included mutual wing-flapping and biting.

HANGING

Keller (1976) defines this action as letting go with the feet and hanging by the bill. In the field we observed keas hanging upside-down from a branch by the feet during social play. The play partner commonly attempted to bite the hanging bird's feet.

JUMPING AND FLAPPING

Keller (1976) describes a behavior he called *Tanzen* (dancing) that occurred during courtship: one bird hops in place, with its wings slightly extended, while looking at the partner, who is at most 30 cm away. In the field we observed keas hopping in place while facing each other and flapping their wings. This behavior occurred in both tussle play and toss play, in association with pushing with the feet, locking bills, and biting. Also included in this category is "jumping on," when one kea approaches another that is rolling over and jumps on its breast (K. J. Potts 1969).

ROLLING OVER

The kea rolls over and lies on its back while gently moving its feet (K. J. Potts 1969). Keller (1975) gives a similar description: the bird rolls over on its back and briefly waves its feet in the air, while making squealing vocalizations. We

TABLE B6 *(continued)*

observed this behavior in response to an aggressive attack (in which the bird immediately rolled over to right itself) and also in high-intensity play. It is associated with jumping and flapping, biting, and pushing with the feet.

TOSSING

The kea first holds an object in its bill and then jerks its head with a vertical movement to release the object in the air; the bird may also hop or flap its wings just before releasing the object (K. J. Potts 1969). It may persist in tossing the object for several minutes. Potts observed young birds repeatedly tossing up the end of a broomstick. We observed keas tossing rocks and small objects, primarily in interactions between adults of opposite sexes, during which the tossing bird would also vocalize and/or jump and flap.

TABLE B7. Sequence of Tussle Play
between a Subadult Male and a Juvenile Male

Subadult Male	Juvenile Male
approach, bite →	*roll over,* stand, jump and flap →
bite, push with foot, *roll over* →	*roll over* →
roll over →	preen self →
approach, *jump and flap* →	*jump and flap* →
roll over →	*roll over* →
jump and flap →	*jump and flap* →
roll over →	jump and flap →
look at partner, preen self →	*look at partner, preen self* →
bite →	roll over →
jump and flap, bite, push with foot →	push with foot, jump and flap, bite, *roll over* →
walk away, return, *roll over, push with foot* →	jump and flap, bite, *push with foot* →
walk away →	preen self, look at partner →
approach, *jump and flap, bite, push with foot* →	jump and flap, bite, push with foot, *roll over* →
roll over, bite, push with foot →	bite, *roll over* →
roll over, bite →	*roll over, bite,* push with foot, run away →
approach, *jump and flap* →	*jump and flap* →
approach →	walk away →
approach, *jump and flap* →	*jump and flap,* walk away →
approach quickly →	run away →
look at partner →	fly away

(continued)

TABLE B7 *(continued)*

NOTE: We recorded this sequence at the Arthur's Pass National Park study site on January 8, 1991, at 6:54 A.M.; it lasted 2 minutes and 39 seconds. Italics highlight social actions that immediately precede or follow identical ones by the play partner, revealing the high degree of facilitation in spontaneous social play. Of the twenty behavioral exchanges in this bout, fourteen included a mirroring of the partner's immediately preceding behavior pattern. This sequence also illustrates the high frequency of play behaviors during an interaction. The major actions, on the part of the subadult and the juvenile, respectively, were biting (7 and 5 times), pushing with the foot (5 and 4 times), rolling over (7 and 8 times), and jumping and flapping their wings (6 and 9 times).

TABLE B8. Sequence of Tussle Play
between a Fledgling Male and a Subadult Female

Fledgling Male	Subadult Female
approach, try to take object* →	push with foot →
take object*, approach →	walk away →
approach, *lock bills* →	*lock bills* →
push with foot, jump and flap →	*jump and flap, push with foot, lock bills* →
lock bills →	look at partner →
approach, *jump and flap, push with foot* →	*jump and flap* →
jump and flap →	walk away →
approach, bite, *jump and flap* →	walk away, *approach,* crouch, *jump and flap* →
jump and flap →	*look at partner* →
look at partner →	approach, *jump and flap, vocalize* →
jump and flap, vocalize →	roll over, *vocalize, lock bills* →
lock bills, jump and flap →	walk away, preen self, look at partner, vocalize, approach, *jump and flap* →
push with foot →	jump and flap, walk away, rummage →
approach →	bite →
jump and flap →	vocalize →
approach →	run away, vocalize, rummage, crouch, approach, crouch, jump and flap →
pick up object, hold in bill →	vocalize

(*continued*)

TABLE B8 *(continued)*

NOTE: We recorded this sequence at the Arthur's Pass National Park study site on January 20, 1990, at 7:06 A.M.; it lasted 3 minutes and 39 seconds. Italics highlight social actions that immediately precede or follow identical ones by the play partner. This sequence is less intense than the interaction diagrammed in table B7. The primary play behaviors, on the part of the fledgling male and the subadult female, respectively, were locking bills (3 times each), biting (once each), pushing with the feet (3 and 2 times), rolling over (0 and 1 time), and jumping and flapping their wings (8 and 7 times). On three occasions the subadult female crouched at the approach of the fledgling male. It is not unusual to see submissive behavior by females toward males, even ones that are much younger.

*The item taken was a piece of paper. In this case the fledgling was stealing a toy rather than food.

TABLE B9. Sequence of Toss Play
between an Adult Male and an Adult Female

Adult Male	Adult Female
approach →	crouch, look at partner →
bite →	walk away →
move slowly →	grasp object, pull it, move slowly, approach →
scrape object →	move slowly, approach →
pick up object, carry it →	approach, touch object with bill, pick it up, toss →
carry object, *jump and flap* →	*jump and flap,* pick up object, hold it in bill →
hold object in bill, toss →	hold object in bill →
scrape object →	toss →
jump and flap →	*jump and flap,* scrape object →
jump and flap →	*jump and flap* →
jump and flap, walk away →	move slowly, touch object with bill, grasp it, hold it down with foot, pick it up, toss, grasp object, toss, grasp object, toss →
eat →	approach, look at partner →
eat →	scrape object, hold it down with foot, scrape object, hold it down with foot, scrape object, hold it down with foot, scrape object, pick up object, toss, fly, move slowly, touch object with bill, move slowly, grasp object, toss, grasp object, toss, jump and flap, toss →
hold object in bill, pry object →	attempt to take object →
push with foot, approach →	walk away, hold object in bill, rummage

(continued)

TABLE B9 *(continued)*

NOTE: We recorded this sequence at the Arthur's Pass National Park study site on January 15, 1990, at 8:32 P.M.; it lasted 5 minutes and 33 seconds. Italics highlight social actions that immediately precede or follow identical ones by the play partner. Each occurrence of "toss" is underlined. In this sequence the female tries to interest the male by repeatedly tossing small stones. The male, however, loses interest and begins foraging. Note that the female tosses the stones at two critical points in the sequence: first, right before the partners engage in facilitated "jump and flap" behaviors; and second, when the male starts to lose interest. Even repeated tosses, however, did not entice the male to resume playing. In this sequence the only play behavior aside from tossing was jumping and flapping. It is not unusual for episodes of toss play to be less intense than those of tussle play and to be interspersed with episodes of solitary foraging.

TABLE B10. Sequence of Tussle Play between Two Fledglings

Fledgling Female	Fledgling Male
approach, *pull object* →	*pull object* →
bite, push with foot →	*bite*, <u>attack</u> →
lock bills →	*lock bills, push with foot* →
bite, *push with foot*, <u>attack</u> →	jump and flap →
roll over →	*roll over* →
walk away →	roll over →
approach, push with foot, look at partner →	roll over →
approach, *scrape object* →	*scrape object* →
push with foot, attempt to take object →	<u>attack</u>, take object, jump and flap →
run away →	hold object down with foot, scrape it, pick up object, carry it, hold it down with foot, scrape it →
approach, bite →	push with foot →
bite →	roll over, pick up object, hold it in bill →
bite, attempt to take object →	roll over, hold object in bill →
pry object →	take object →
attempt to take object →	approach, grasp object, scrape it, run away

NOTE: We recorded this sequence at the Arthur's Pass National Park study site on January 13, 1990, at 7:38 A.M.; it lasted 2 minutes and 53 seconds. Italics highlight action patterns that immediately precede or follow identical ones by the playing partner. Play interactions between fledglings show very high levels of aggression; serious attacks (underlined) are interspersed with more characteristic play behaviors. Note also the facilitation of demolition behaviors (pulling and scraping) in the context of a play interaction.

TABLE B11. Effect of Social Dominance on Hunching

Dominance Category	Number in Category	Observed Number of Hunches	Expected Number of Hunches	Chi-Squared
		Adult Recipients of Juvenile Hunches		
−3	8	17	24.216	3.063
−2	6	21	18.162	0.383
−1	6	16	18.162	0.292
1	4	16	12.108	0.947
2	9	18	27.243	4.747
3	4	24	12.108	5.892
Total	37	112	112	15.324*
		Juveniles Hunching to Adults		
−3	2	7	7.789	0.089
−2	6	22	23.368	0.085
−1	3	15	11.684	0.733
1	5	17	19.474	0.360
2	2	10	7.789	0.489
3	1	3	3.895	0.267
Total	19	74	74	2.023†

NOTE: We estimated the social status of each banded individual in terms of a dominance index. The index is the number of interactions in which a given bird displaced another, minus the number in which it was displaced, divided by the total number of its interactions. The index is positive for birds that tend to displace others and negative for those that tend to be displaced. For this analysis, we grouped the dominance indices into six categories:

−3 = indices of less than −0.50
−2 = indices between −0.50 and −0.25
−1 = indices between −0.25 and 0.00
 1 = indices between 0.00 and 0.25
 2 = indices between 0.25 and 0.50
 3 = indices between 0.50 and 0.75.

We obtained the dominance category for each juvenile and adult who participated in a hunch interaction and tested to see whether the dominance status of either the displaying or the recipient individual significantly affected the incidence of hunching.

(continued)

TABLE B I I *(continued)*

*The data reveal a significant difference between the distribution of hunches to adults and the distribution of adult dominance scores. Juveniles appear to hunch more often than would be expected to low-status adults and less often than expected to adults in the highest dominance category. This observation supports the notion that hunching is partly an aggressive display; if it were simply submissive, it would be displayed more rather than less often to highly dominant males.

†The dominance status of juvenile males does not seem to influence their tendency to produce hunches: the distribution of juvenile hunches does not differ significantly from the distribution of juvenile dominance. Low-status juveniles are as likely as high-status ones to display hunches to adults. This observation supports the notion that hunching is an appeasement display overlaying the normal aggressive behaviors involved in dominating and displacing other birds.

NOTES

INTRODUCTION

1. J. M. Diamond 1990b, 3.

CHAPTER ONE

1. J. M. Diamond 1990a.
2. Godly 1975; McGlone 1988; Allen and Platt 1990.
3. A. Anderson 1984; Daugherty, Gibbs, and Hitchmough 1993; Worthy and Holdaway 1993; Worthy 1989. The tallest moa was *Dinornis giganteus,* roughly as tall as an ostrich, though far more massive (Brewster 1987, 5). There was much confusion in the early taxonomy of the moa. The group was originally divided into as many as thirty-seven species, whereas only eleven are recognized today (A. Cooper, Atkinson, Lee, and Worthy 1993). Moa species names are listed in table 2 (see chapter 2).

4. A. Anderson 1982; A. Cooper, Mourer-Chauviré, Chambers, von Haeseler, Wilson, and Pääbo 1992; Burrows 1980; Campbell 1976, 79; Riney, Watson, Bassett, Turbott, and Howard 1959, 39–40. Greenwood and Atkinson (1977) suggest that various plants developed specialized adaptations to grazing by moas, including divaricating foliage, distasteful secondary compounds, and tough fibrous leaves.

5. On takahes, see Beauchamp and Worthy 1988; Lavers and Mills 1984, 1–24. On kakapos, see Worthy and Holdaway 1993; H. Best and Powlesland 1985, 3–32. The takahe was considered extinct until a small population was rediscovered in 1948 in the mountains of Fiordland, in the extreme southwest of the South Island. Kakapos are now among the rarest species of birds, found only on a few offshore islands in New Zealand. Intensive recovery programs for both takahes and kakapos are currently underway.

6. R. N. Holdaway 1989a; on wetas, see O'Brien 1981, 64. The largest forest wetas, though common in prehuman times, now occur only on offshore islands that are free of rats (O'Brien 1981, 113). On New Zealand wrens, see Heather and Robertson 1997, 371. New Zealand wrens constitute the endemic family Acanthisittidae. The reptilian order Rhynchocephalia, to which the tuatara belongs, is otherwise known only in fossils from the late Jurassic (Fleming 1979, 30).

7. R. N. Holdaway 1989a. On the huia, see Burton 1974; Moorhouse 1995, 82–102; Phillipps 1963, 24. On piopios see Heather and Robertson 1997, 422; on owlet-nightjars, Feduccia 1996, 323.

8. Craig 1985; Gaze and Clout 1983; Bergquist 1985; McEwen 1978.

9. On moreporks, see Moorhouse 1995, 5; on adzebills, R. N. Holdaway 1989a.

10. Millener 1990; R. N. Holdaway 1989b; Oliver 1930, 392–94; Bull and Whitaker 1975. There is no consensus on the mode of foraging of the Haast's eagle. Trotter and McCulloch (1984, 50) argue that *Harpagornis* may have been a carrion eater with diminished powers of flight.

11. R. N. Holdaway 1989a.

12. R. N. Holdaway 1989a; Worthy and Mildenhall 1989.

13. Jackson 1960; McGlone 1989; Allen and Platt 1990. Masting is much more pronounced at higher altitudes, and mountain beech is one of the most variable mast-seeding species in New Zealand (Webb and Kelly 1993).

14. Brejaart 1988, 37–41; Riney et al. 1959, 39–40; Campbell 1976, 15–26; Jackson 1960; Guest 1975; Clarke 1970; Myers 1924; Jackson 1969. Similar seasonal variations in kea foraging have been recorded from Routeburn Basin in Fiordland (Campbell 1976, 15–26) and Craigieburn Forest Park (Brejaart 1988, 58).

15. Riney et al. 1959, 40; Campbell 1976, 42, 71; White 1894; Aspinall 1990, 6. Keas have been known to hunt and kill small animals, at least in captivity. For example, young keas hatched in the Zurich Zoo captured and ate sparrows (Yealland 1941). Keas have been seen catching live mice and are even said to have killed full-sized rats by nipping them behind the neck (S. Porter 1947). Interviews with zookeepers reveal that captive keas have killed and eaten larger prey as well, including wekas (Wayne Schulenburg, personal communication, 1988). One rather gruesome anecdote even suggests cannibalism. The keeper at a zoo with a large kea aviary remarked that no matter how many new birds were introduced, the numbers seemed to remain about the same, and that scattered body parts were occasionally found in the cage. Although some of these birds must have eaten their companions, it was not clear whether they had killed them outright or had simply scavenged their remains after the birds had succumbed to other causes.

16. Brejaart 1988, 37–42; Campbell 1976, 15–26.

17. McGlone 1989.

18. On moas, see Trotter and McCulloch 1984, 50–53; A. Anderson 1989; Brewster 1987, 14–15. On the basis of midden counts and present-day emu densities, Anderson estimates that there were between 37,000 and 100,000 moas on the South Island between four hundred and one thousand years ago. Brewster quotes a figure of thirty moas per square kilometer of forest, but it is difficult to extrapolate this value to the is-

land as a whole. On kakas, see IUCN/SSC Captive Breeding Specialist Group 1993; O'Donnell and Dilks 1994; on kakapos and takahes, Worthy and Mildenhall 1989.

19. Millener 1990; Sprecht, Dettmann, and Jarzen 1992.
20. Gibb 1990; O'Brien 1981, 11–12.
21. Forster 1975; R. A. Cooper and Millener 1993; A. Cooper et al. 1992; Millener 1990; A. Cooper et al. 1993. The notion that the moas' ancestry dates to the breakup of Gondwanaland is common in the literature (e.g., Cracraft 1980), but it may derive from a failure to realize that the ratites (wingless birds) are undoubtedly a polyphyletic group with multiple, independent origins. On the basis of a variety of highly distinctive morphological features, Feduccia (1996, 278) argues that the moas' ancestors probably flew to New Zealand after it was isolated from the rest of the southern continents.
22. Millener 1990; Daugherty et al. 1993; McGlone 1988.
23. R. N. Holdaway 1989a; Webb and Kelly 1993. Both species of bats are endemic to New Zealand, but the short-tailed bat is the sole representative of an endemic family. Behaving much like a shrew or a honey possum, it exploits an unusually diverse range of foods. It feeds on the ground, on foliage, and on tree trunks; it takes insects in flight; and it also eats fruits and nectar. This extraordinary creature even crawls along the ground and pollinates a nonphotosynthetic flowering plant that parasitizes the roots of trees and shrubs.
24. R. A. Cooper and Millener 1993.
25. As a whole, parrots are an ancient lineage, only distantly related to other birds. Parrots from Southeast Asia, Australia, and the Pacific appear to form a distinct phylogenetic grouping (Fleming 1962; Forshaw 1977, 19–21; Kavanau 1987; Christidis, Schodde, Shaw, and Maynes 1991). Although Sibley and Ahlquist's monolithic study of avian DNA has validated the taxonomic distinctiveness of these Australo-Papuan parrots (1990, 389–90), their sample of species was inadequate to make finer differentiations. Van Dongen and De Boer's karyotype

studies (1984), which included the kea, confirm earlier suggestions that the parrots' pattern of evolution is widely variable and complex but do not allow the authors to define the origins of the kea with any confidence. Taxonomies based on morphology have produced vastly differing accounts of affinities within the group. According to Holyoak (1973), *Nestor* evolved from the Loriinae, the Indonesian lories and lorikeets, and migrated into New Zealand from the north. G. A. Smith (1975), however, places *Nestor* in a tribe most closely related to *Strigops,* the New Zealand kakapo, and less closely related to the Platycercini and the Cacatuini, the Australian cockatoos.

26. Fleming 1975; 1979, 89–90.
27. R. A. Cooper and Millener 1993; McGlone 1988.
28. Mayr 1966. Cracraft (1980) discusses the same processes with respect to the diversification of moa species.
29. Fleming 1979, 89–90; 1975. Fleming (1980) notes that his account cannot explain why keas, if they evolved earlier in the Pleistocene, did not extend into the North Island across the land bridge in the last glaciation. One possibility is that speciation dated from the last (i.e., postglacial) flooding of the Cook Strait, but that seems far too recent, considering the degree of differentiation that has occurred. Later research (R. N. Holdaway and Worthy 1993), however, indicates that the species may in fact have extended north: fossil kea remains have been identified in North Island deposits dating from the last Otiran glaciation, ten thousand years ago. This finding is not surprising, since sea levels were lower during the glacial epochs, and the two islands were joined into one. Although none of the area that is now the North Island was actually glaciated, tundra and alpine scrub habitat extended far north of the land bridge at the maximum extent of the ice, and forests were restricted mainly to coastal areas (see map in Fleming 1975, 53).
30. McGlone 1988; Fleming 1979.
31. McGlone 1988.
32. Worthy and Holdaway 1993.

33. When two competing species coexist on an island, one of the species is often greatly reduced in numbers or lives only in strongly demarcated clumps (R. A. Wallace 1978). We suggest that both these factors may have been at work when kakas reinvaded the South Island around nine thousand years ago: keas may have been forced to live as a relict species in marginal patches of alpine and forest habitat.

CHAPTER TWO

1. Trotter and McCulloch 1989, 24–25; Caughley 1988; Leach 1969, 29, 74.
2. Green 1975; Cowen 1905; R. Moorhouse, personal communication, 1998.
3. Davidson 1984, 134; Orbell 1985, 21, 24; Phillipps 1966, 33–34. The Maori saying is from E. Best 1977, 193.
4. E. Best 1977, 192–216; Beattie 1994, 344. Temple (1996, 56) describes a legend of the Waitaha Maori in which keas were considered *kaitiaki*, or guardian birds, for tribe members who traveled across the mountains to the sources of greenstone.
5. Arthur's Pass National Park 1986, 60–63; Leach 1969, 72. Trotter and McCulloch (1989, 86–87) dispute the standard account of the movements of the early Maori. They do not believe that the Maori had the technology to work greenstone until about four hundred years ago and question whether they would have sought out and transported the raw material before then. Morris and Smith (1988, 62) suggest that the early Maori may not have spent much time in the Southern Alps because their maritime culture led them to regard the mountains as *tapu* (taboo). If these views are valid, the Maori may not have encountered keas until perhaps the sixteenth century; this late discovery would then account for the limited role keas played in Maori culture.
6. Cassels 1984; Scarlett 1979; A. Anderson, McGovern-Wilson, and Holdaway 1991; Millener 1990.
7. S. Holdaway 1991.

8. Andrews 1986, 11, 14, 25; Medway 1976; Reed and Reed 1969, 157–221, 244.
9. Turbott 1967, 87; Andrews 1986, 58–61.
10. Andrews 1986, 99.
11. K. J. Potts 1976; Oliver 1930, 407; Gould 1856; Turbott 1967, 92.
12. Buller 1869, 1870; Travers 1883. The quotes are from Turbott 1967, 94.
13. Buller 1883; Myers 1924; Turbott 1967, 94.
14. Baggaley 1967, 75–76; Watters 1965, 52.
15. Watters 1965, 53–54; Levy 1970, 239–47.
16. Watters 1965, 59; Dadelszen 1904, 343–44.
17. Turbott 1967, 95.
18. T. H. Potts 1871; Marriner 1908, 72–75.
19. Marriner 1908, 138.
20. Menzies 1878; T. Kirk 1895.
21. Marriner 1908, 138–39; W. W. Smith 1888; Oliver 1930, 407; idem 1955, 542; Falla, Sibson, and Turbott 1978, 163; Myers 1924. R. N. Holdaway and Worthy (1993) dispute the notion that the kea's range greatly expanded during the 1800s. They suggest that keas were always in the northern part of the South Island, but by the 1870s there were more observers in the area, giving rise to more reported observations. Because settlement in the mountains of the South Island began in the south and only gradually expanded northward, it is difficult to infer the overall distribution of keas based on sightings by early explorers. Marriner (1908, 136–39) interprets much of the apparent range extension before 1880 as an illusion due to the patterns of settlement, but he notes that after 1880 the species was indeed sighted in previously settled habitats in which they had never been seen, presumably extending their range from the high mountain areas as their populations increased.

R. N. Holdaway and Worthy (1993) cite records of fossil keas found in caves in northern Nelson, as well as their discovery of kea fossils on the North Island, to suggest that if the kea population actually did expand northward, the birds were simply recolonizing an area they had previously inhabited. It is possible that the original range of the kea did include the southern portion of the North Island, at the time when the North and South Islands were united into one landmass during the last glaciation. If so, keas on the North Island would have died out sometime after the islands separated, and the species may have retreated from Nelson after the expansion of the forests brought back kakas and other competitors. Reports of subfossil remains of keas on more remote New Zealand islands (e.g., E. Dawson 1959) seem rather unlikely and have not been substantiated. Today, while keas are still often seen in subalpine forests in Nelson and along the west coast of the South Island (O'Donnell and Dilks 1986), the only keas found on the North Island seem to be ones that have escaped from captivity (A. Cunningham 1974).

22. Salmon 1975.

23. Acland 1975, 178–79.

24. Turbott 1967, 96. Given the current exchange rate, adjusted for inflation, the bounty would translate to about forty U.S. dollars per kea—a truly impressive sum in a depressed rural economy (Temple 1994).

25. Marriner 1908, 123, 128–32.

26. Turbott 1967, 95–96; Marriner 1906, 1907; idem 1908, 122. White 1894 provides additional descriptions of kea attacks on sheep. Jackson (1962a) assessed the credibility of Marriner's informants and concluded that eight of the fourteen accounts in Marriner 1908 contained false statements. He found it plausible that keas would attack sheep that were trapped in snow, were sick or injured, or appeared to be dead, but he did not believe the birds posed any risk to healthy, mobile livestock.

27. Binney 1990, 107.

28. T. H. Potts 1871. Potts's account captured the attention of one of England's premier naturalists, Alfred Russel Wallace, who used the kea as an example in his book-length defense of the theory of natural

selection (A. R. Wallace 1891, 75). Buller, skeptical of Wallace's knowledge of the New Zealand fauna, claimed that Wallace had originally mistaken kakapos for keas: "I may refer to a very curious mistake made by Mr. Alfred Russel Wallace, the great apostle of the creed of natural selection, to whom, indeed, we all metaphorically doff our hats in respectful admiration. In writing of the New Zealand avifauna he confounds the Kakapo with the Kea, declaring that the moss-eating Stringops [sic] had become carnivorous and is most destructive to the settlers' sheep" (Buller 1894, 92).

29. T. H. Potts 1871.
30. Benham 1906, 83.
31. Menzies 1878, 376–77.
32. Marriner 1908, 98–101; Benham 1906. See also the maggot theory (Reischek 1885).
33. Marriner 1908, 100.
34. McKay 1884.
35. Turbott 1967, 94.
36. T. H. Potts 1870, 88.
37. Benham 1906, 82; Buller 1883, 316.
38. Between A.D. 900 and 1840 the New Zealand forest was reduced from 78 percent to 53 percent of the land area. European settlers subsequently brought the total forest acreage down to 23 percent (Atkinson and Cameron 1993).
39. King 1984, 65–81; J. M. Diamond and Veitch 1981; Atkinson and Cameron 1993; Wodzicki and Wright 1984.
40. Clark 1970, 259–82; Gibb 1990; King 1984, 112–14. Brush-tailed possums are particularly destructive, because they eat young shoots and fresh leaves of many native trees that form the forest canopy. Moreover, both red deer and possums browse on beech seedlings, reducing the recovery of the canopy and eventually opening the forests, thus making it even easier for deer to browse (Salmon 1975).
41. Atkinson and Cameron 1993; J. M. Diamond and Veitch 1981.

42. Atkinson and Cameron 1993; King 1984, 72–74; R. N. Holdaway 1989a; R. Moorhouse, personal communication, 1998.

43. Subfossil remains show that at least thirty-four native nonmarine bird species became extinct during the Maori settlement before Europeans arrived (J. M. Diamond and Veitch 1981). Since 1981, two more extinct species of New Zealand wrens have been described, both probably present at the time New Zealand was first settled (Millener 1988). Of the seventy-seven native nonmarine bird species present when Europeans arrived in the early nineteenth century, eight have become extinct and thirteen have become rare or local (J. M. Diamond and Veitch 1981). See also Steadman 1995; Ornithological Society of New Zealand 1970, 77–79; A. Cooper et al. 1993; R. N. Holdaway 1989a; Bull and Whitaker 1975; Millener 1990; A. Cooper et al. 1992.

New Zealand and the Philippines have over twice the percentage of threatened birds as any other country (Baillie and Groombridge 1996, intro. 33). Because of their initially lower population densities and low resistance to introduced predators and competitors, insular species are at a particularly high risk of extinction. They constitute 90 percent of the bird species that have become extinct worldwide during historic times; not surprisingly, then, most of the currently endangered bird species live on islands (Snyder, Wiley, and Kepler 1987, 6).

Since 1840 three plant species endemic to New Zealand have apparently become extinct and forty-five have become highly threatened. No plants are known to have become extinct as a consequence of Polynesian settlement (Atkinson and Cameron 1993).

44. Worthy and Holdaway 1993. King (1984, 190) states that Australasian harriers have also benefited from the changes people brought and have become more abundant in New Zealand, although Heather and Robertson (1997, 276) claim that their numbers have declined since the 1950s because of rabbit control, which has removed a large part of their diet. The harrier has a transcontinental range, however, including Australia and much of the southwestern Pacific (Heather and Robertson 1997, 275–76).

CHAPTER THREE

1. Jackson 1962b; Arthur's Pass National Park Board 1958. Observations we made at Fox Glacier in Westland National Park supplement our field study. The site is a glacial moraine at the foot of Mount Tasman, one of New Zealand's highest peaks. The glacier has been retreating and advancing since about A.D. 1200. Its most recent advance was 180 m, between 1965 and 1968. A road runs up the valley along rock debris to the terminal moraine and ends at a small parking lot. Keas live in the surrounding shrub land and frequent the parking lot whenever people are present. The habitat consists of dense stands of native broom associated with willow-leaved hebe, tree daisy, and tree tutu (Westland National Park 1982, 1–12).

2. Jackson 1960, 1962a, 1962b, 1963a, 1969.

3. Peter Simpson, personal communication, 1990.

4. Altitudes range from 270 to 2,400 m, with timberline at about 1,200 m (New Zealand Department of Conservation 1989, 3).

5. N.Z. Dept. of Conservation 1989, 8; Salmon 1985.

6. Yearly rainfall averages about 450 cm, and as much as 25 cm may fall in the course of a single day (N.Z. Dept. of Conservation 1989, 6–7).

7. Arthur's Pass National Park 1986, 123; N.Z. Dept. of Conservation 1989, 6.

8. On mountain beech, see Stewart 1984, 88; McGlone 1988.

9. Keller 1972; Lint 1958; Mallet 1973; Schmidt 1971.

10. Bond, Wilson, and Diamond 1991.

11. We have analyzed material from this database in J. Diamond and Bond 1991; Bond and Diamond 1992.

12. See appendix table B1.

13. On the versatility of the kea's bill, see English 1939. Keas often dig with their upper bills, drawing a furrow in the soil or probing to a depth of several centimeters. The only other parrot with a similar bill is the long-billed corella, which also spends a lot of time digging roots (Forshaw 1977, 135). A kea can use its bill laterally as a spade to remove

unwanted soil and to flick aside snow. The birds often remove the top layer of leaf litter before beginning to dig, flipping larger pieces away with their bills. In the San Diego Zoo we observed several keas raking litter aside with a twig.

Because of its narrow cross-section, the upper bill lends itself to prying and probing. It can be drawn down a narrow crack or forced into a small hole to extract concealed insects and other food. A kea will often wedge its upper bill into an opening, then lever it to pry apart a rotten log or roll a rock. Keas also hook their bills into soft material, such as an animal carcass, and pull to tear an opening. The upper bill provides a fulcrum for scraping with the chisellike lower bill to remove fragments of dried meat from bones, flesh from soft fruits, or bark from tree limbs (Keller 1976). Joining the upper and lower bills like a pair of long-nosed pliers, keas twist and break off pieces of food or tear open animal skin. The remarkable musculature of the parrot jaw, which allows for independent movements of the upper and lower bills, makes the kea's bite highly effective (Forshaw 1977, 26; McCann 1963).

The kea's tongue is thick and muscular, tipped with a fringe of black, hairlike papillae on the outer edge (Garrod 1872; McCann 1963; E. J. Kirk, Powlesland, and Cork 1993; Holyoak 1973). This brushlike structure, which is also found in a variety of other parrots, probably helps in feeding on powdery or liquid food, such as pollen and nectar. Keas drink water by lapping it directly or by scooping it with their lower bills (Keller 1976). They process soft food by rolling it back and forth between the tongue and the upper bill (Brejaart 1988, 50–51), or they tear it apart by pressing it against the hard palate and rasping off particles with the tongue (Keller 1976).

On use of the feet in feeding, see G. A. Smith 1971; Keller 1976. Keas invariably pick up food with the bill and then transfer it to the foot. They do not carry food in their feet during flight.

14. Differences in foraging behaviors between sex and age groups are displayed in appendix table B2. J. Diamond and Bond 1991 provides a full descriptive ethogram of the foraging actions.

15. Bond and Diamond 1995, 1992. The females that do visit the site, however, generally come as often and stay as long as males of the same age (see appendix table B3).

16. Huntingford and Turner 1987, 46–49, 228–48; Archer 1988, 105–58; Rowell 1966.

17. The incidence of attacking declines with age in both males and females (see the interpretation given below appendix table B4). At the very least, female keas appear to be every bit as aggressive as males.

18. Subadult males are generally subordinate and readily displaced by adults; all but one of the thirty adult males in our sample outranked all subadults. Subadult males do, however, fare better against juveniles, whom they outranked in fifteen of the thirty-eight matchups we observed. These results might suggest a linear hierarchy with a fairly strong age dependence, but some juvenile males also ranked higher than some adults. Of the six juvenile males that were outranked at least some of the time by subadults in our sample, all ranked higher than at least two or three adult males, and one ranked higher than seven of the ten banded adults recorded in his field season. See appendix table B5 for more detail. A study of a captive group of seven keas by Tebbich, Taborsky, and Winkler (1996) also clearly shows a nonlinear dominance structure. Three of their seven birds had at least one relationship in which dominance rank was reversed from a strictly linear pecking order.

Females of all ages are uniformly outranked by adult and subadult males, although their rankings relative to fledgling and juvenile males are more equable. Among females there is a significant ordering of dominance by age, similar to the pattern among males. Adult females are more dominant than juvenile and subadult females, and female fledglings are also more dominant than subadult females (appendix table B5). Subadult females seem to be under the same general social disadvantage as subadult males in terms of displacement from food resources.

19. Heinrich 1989, 200, 201–11, 229.

20. Child 1975; Kikkawa 1966; Arthur's Pass National Park 1986, 116–17.

21. The New Zealand falcon was probably originally distributed widely throughout New Zealand (Soper 1965, 42). Forest clearing and hunting have reduced the population to only about four thousand pairs. They are most common on the North Island and the northwestern parts of the South Island (Heather and Robertson 1997, 277).

22. The earliest record is T. H. Potts 1870. See also R. E. R. Porter and Dawson 1968 and Jackson 1969.

23. Jackson 1969.

24. See appendix table B3. See also Belcham 1988, 48; Campbell 1976, 11; J. Diamond and Bond 1991.

25. On panting in other birds, see Bartholomew, Lasiewski, and Crawford 1968. On thermal stress in parrots, see McNab and Salisbury 1995.

26. K. J. Potts 1969, 25, 31–37; 1976; 1977.

27. Hinde 1974, 8, 332.

28. Skutch 1976, 266–72. The time course for regurgitant feeding may be shortened in captivity. Zeigler (1975) noted that parental feeding declined sharply about four weeks after fledging in a family group of captive keas. A low level of regurgitant feeding continued, however, through the eighth week. Three hand-reared chicks born at the San Diego Zoo in 1979 were successfully weaned about fifty days after hatching (J. Mitchell 1981).

Raptors also have an extended period of regurgitant feeding. Johnson (1986) remarked that red-tailed hawks continue to provide food to their young for at least fifty-three days past fledging. As in keas, development of self-sufficiency in red-tails is a gradual process, in which juveniles progressively capture more and more of their own food while parents provide less.

29. See appendix table B4. Adult males show significantly lower levels of crouching than adult females and lower levels than any of the younger males, in particular subadults.

30. Forshaw 1977, 30.

31. Play has been observed in sixty-two species of birds in fourteen

different orders. Extensive object and social play, however, is apparently restricted to parrots, corvids, hornbills, and certain woodpeckers. See Deckert and Deckert 1982; Fagen 1981; Ficken 1977; Gaston 1977; Gwinner 1966; Jackson 1963b; Keller 1975; Moreau and Moreau 1944; Müller-Schwarze 1978; Ortega and Bekoff 1987; Pellis 1981, 1982; Skeate 1985.

Naturalists have remarked on the play of keas since before the turn of the century. Marriner (1908, 60–71) devotes a full chapter to anecdotes of keas playing with objects. Keas exhibit similar behavior in captivity, in spite of the depauperate environment of zoo aviaries. In 1947 Derscheid noted that his captive keas "play in the most unexpected way with the most odd things. They occupy themselves, for example, in emptying the water out of their bath, using a cup or metal box for this purpose. Snow is a source of amusement for them, and I have often seen them playing by fishing out the pieces of ice from the edge of a little stream that flowed through their aviary" (Derscheid 1947, 49). J. Diamond and Bond (1989a) documented extensive water play in a group of captive keas at the San Diego Zoo. Kea social play, unlike their object play, has been described only in captivity (Keller 1975, K. J. Potts 1969, 39–45). Neither type of play in keas has ever been quantitatively analyzed before.

32. See appendix table B6. Although it is easily recognized, play is surprisingly difficult to define. Researchers agree on various features as indicative of play (Bekoff 1972, 1976, 1984, 1989; Fagen 1981, 42–54; idem 1982; Martin and Caro 1985; R. W. Mitchell 1991; Müller-Schwarze 1978, 1–4; Ortega and Bekoff 1987). No single criterion is either necessary or sufficient to define a play episode, however, and the indicative features may or may not be evident in any given play sequence.

33. Jackson (1969) observed a young kea engage in this behavior while playing with its mother. The juvenile male swooped down on his mother as if he were a falcon, then watched as the tables were turned and the mother swooped down on him. He rolled on to his back and

prepared to parry the blow from his mother's play attack with his feet. The behavior is also described in Jackson 1962b; Keller 1975; and K. J. Potts 1969, 41.

34. Kilham 1989, 69.

35. R. W. Mitchell 1991; Fagen 1981, 414–18; Bekoff 1972.

36. See appendix table B7.

37. On tussle play between male fledglings and subadult females, see appendix table B8. Among males, tussle play is mainly the province of younger birds. Female keas, on the other hand, show elevated levels of tussle play as fledglings, subadults, and adults (see appendix table B4).

38. On the toss play sequence, see appendix table B9. On the frequency of toss play among different groups, see appendix table B4.

39. Fagen 1981, 414–18.

40. On tug-of-war and keep-away, see Fagen 1981, 117, 140; R. W. Mitchell and Thompson 1986. Kubat (1992, 40–41) noted that captive keas display the same dominance ordering in obtaining access to food as in obtaining access to play objects.

41. Loveland 1986 provides an example of the term "affordances" to describe the properties and potential uses of an object; the original coinage is from Gibson 1966, 285. Kubat (1992, 97–101) noted that keas investigate objects that are readily manipulated or dismantled more intensively than ones that are not. If an object was highly attractive, it made little difference whether it was new or familiar (see also Ritzmeier 1995, 25).

42. Ritzmeier 1995, 25; Kubat 1992, 97–101. Keas effectively never tire of certain objects. Twelve months after a mirror had been introduced into its cage, a kea at the San Diego Zoo continued to interact with his mirror image (J. Diamond and Bond 1989b).

43. See appendix table B3.

44. On the age distribution of keas at the site after dark, see appendix table B3. On the nocturnal movements of keas, see Belcham 1988, 50.

CHAPTER FOUR

1. S. Porter 1947; Lint 1958; Schmidt 1971; Jackson 1963a; Schifter 1965.
2. IUCN/SSC 1993, 6, 10; Forshaw 1977, 139; Jackson 1962b, 1963a. In captivity the time to maturity may be longer. Sieber's (1983) captive female was four years old when her first clutch was laid.
3. See appendix table B2, notes on scraping and gleaning.
4. Ficken 1977; Higuchi 1986; Skutch 1976, 321.
5. J. Diamond and Bond 1991.
6. On the incidence of aggression and play by age and sex, see appendix table B4. For the breakdown of one play sequence between fledglings, see appendix table B10.
7. K. J. Potts 1977.
8. See appendix tables B2 and B4.
9. Juvenile appeasement displays are also known in other birds. Juvenile herring gulls, for example, can inhibit adult aggression by a display coincidentally referred to as "hunching" (Drury and Smith 1968; Tinbergen 1960b, 182–83). A similarly favored social status for juveniles that is not evidently mediated by appeasement has been recorded in several species of jays (Balda and Balda 1978; Barkan, Craig, Strahl, Stewart, and Brown 1986; Brown 1963; Brown and Brown 1984; Lockie 1956).
10. See appendix table B4, note on hunching.
11. J. Diamond and Bond 1991.
12. On appeasement displays, see Tinbergen 1973; Hinde 1981. For additional detail on the relationship between dominance and hunching in keas, see appendix table B11.
13. Stephen Phillipson, personal communication, 1991, notes kea sightings at the airport; Jackson 1960 describes juvenile aggregations.
14. J. Diamond and Bond 1991. These observations contrast with studies of other bird species, which find that foraging efficiency increases monotonically with age (e.g., MacLean 1986, Burger 1986).

15. J. Diamond and Bond 1991.

16. Stealing of food occurs significantly more frequently among subadult and adult keas (pooled across sex) than among juveniles and fledglings (see appendix table B2). This finding obscures the striking dominance of the behavior among subadult males. About half of all adult males and fledglings engaged in theft to some degree. The same proportion held true for females of all ages. More than 70 percent of juvenile males stole food from other individuals, however, and all subadult males did so. Moreover, the frequency of theft among subadult males was uniformly high. This observation leads us to characterize stealing as a quintessential subadult behavior.

The relationship between the age class and the incidence and success of piracy is quite variable in other bird species, even among seabirds, for whom theft is a favored foraging technique. Juvenile herring gulls have been observed to steal more food than adults, but the reverse is true of laughing gulls (Burger and Gochfield 1981). Among kelp gulls, kleptoparasitism decreases with age, while the efficiency of foraging behavior increases (Hockey, Ryan, and Bosman 1989). Burger and Gochfield (1979) found that although subadult ring-billed gulls engaged in more attempts to steal, they were less successful than adults. Brockman and Barnard (1976) note the lack of kleptoparasitism in other parrots.

17. Jackson 1963a.

18. The color pattern of ravens may similarly change with their release from social domination. Heinrich (1989, 144–45) noted that juvenile ravens, which characteristically have pink mouth linings, develop the black mouths of adults more rapidly when they are kept in captivity. Color change in keas is also much faster in captivity. Keepers at both the Auckland Zoo and the San Diego Zoo noted that juvenile keas lost all their yellow color within a year after fledging. Heinrich further remarks that many birds change bill color under hormonal influence as they enter breeding condition, and that breeding condition in birds is highly sensitive to psychological stimuli. He suggests that the domi-

nance structure could be acting through the endocrine system to suppress maturation in subordinate individuals. Juvenile birds brought into captivity are prematurely separated from dominant adults, and their maturation accelerates accordingly.

19. Lorenz 1976.
20. Ibid.
21. Clayton (1978) originated the narrower definition of social facilitation; Zajonc (1965) used the term to refer to the effect that the mere presence of another individual, regardless of what it is doing, has on the first individual—a concept that is perhaps overly broad. Our definition distinguishes social facilitation from "local enhancement," in which an animal is simply attracted to the area where other animals are present (Turner 1965). It further incorporates learning due to what Zentall (1996) calls "stimulus enhancement" and perhaps even "observational conditioning," as well as other factors that do not readily fit into preestablished categories. Drawing such fine distinctions has become a kind of cottage industry in the field of social learning, but their value for empirical field studies is dubious. For support for our empiricist position, see Galef 1996; for a detailed taxonomy of mechanisms of social learning, see Galef 1988 and Zentall 1996.
22. Imitation of a foraging technique, or observational learning of what psychologists refer to as an instrumental response, seems to be fairly uncommon among animals unless deliberately encouraged by a regimen of rewards (e.g., Fragaszy and Visalberghi 1996; Sherry and Galef 1990). For examples of lack of teaching and parental indifference (in primates), see Fragaszy and Visalberghi 1996; Cheney and Seyfarth 1990, 218–27.
23. Young capuchin monkeys display a similar lack of awareness of their parents' feeding techniques (Fragaszy and Visalberghi 1996). Field research on Japanese macaques (Huffman 1996) and black rats (Terkel 1996), however, makes clear that foraging techniques can indeed be socially transmitted.
24. An incident we observed between an unrelated pair of male

keas caged together at the San Diego Zoo illustrates the conflict between social facilitation and social dominance. One of the birds was a juvenile raised in captivity since hatching, the other an adult that had been captured in the wild. When we placed pieces of paper smeared with butter in the cage, the juvenile, who was less fearful of our presence, approached the paper, played with it briefly, and dropped it. Immediately the adult swooped down on the discarded paper and consumed the butter. The juvenile subsequently showed great interest in buttered paper, but only because the adult seemed to value it. He would pick up the paper and manipulate it, licking it tentatively. Eventually he would drop it, and the adult would invariably grab the paper and carry it away to feed on in another part of the cage.

25. Fagen 1981, 19; Geist 1978, 139. The quote is from Fagen 1982, 365.

CHAPTER FIVE

1. Tinbergen 1963; Huxley 1963; Darwin 1859, 153. Contrasting the behavior of closely related species to discover the course of evolutionary changes is one of the chief methodologies advocated by ethologists, and it has been used to good effect in studies of other birds that relate their behavior to features of their ecology and life history (e.g., Faaborg and Bednarz 1990). Konrad Lorenz, in particular, made the comparative method the centerpiece of his approach to understanding behavioral evolution. His classic papers, however, compare only behavior patterns that are substantially or entirely innate, such as courtship and territorial displays. Although well aware of the vast differences between species in the role of learning, Lorenz (1976) illustrated the problem of open-program animals by contrasting completely unrelated organisms, such as ravens and grebes. Our study of the kea is among the first to explore broadly the evolutionary sources of behavioral flexibility by comparing the behavior of closely related species in the field.

2. On the two birds' sizes and colors, see Forshaw 1977, 138–39. On

the identification of fossil remains, see R. N. Holdaway and Worthy 1993.

3. The functional differences between kea and kaka bills were noted very early (T. H. Potts 1870; Shufeldt 1918). On sexual differences between the two species, see Moorhouse, Sibley, Lloyd, and Greene forthcoming; Bond et al. 1991. On their leg and wing bones, see R. N. Holdaway and Worthy 1993.

4. Moorhouse and Greene 1995.

5. The overlap in ranges for keas and kakas is noted in Bull, Gaze, and Robertson 1985, 144, 146; the overlap in habitat preferences in O'Donnell and Dilks 1986. The latter study indicates that kakas spend 27 percent of their time feeding on beech, whereas keas spend 36 percent.

6. On kaka diets, see Beggs 1988; Beggs and Wilson 1991; O'Donnell and Dilks 1989, 1986. Kea diets are discussed in Jackson 1960; R. Brejaart, personal communication, 1989. Kakas on Kapiti Island take their sugar mainly in the form of nectar. During the two peak seasons for flowering trees on Kapiti—July and August, and again from November to January—kakas spend up to 80 percent of their time feeding on nectar and pollen (Moorhouse 1995, 47).

7. Beggs and Wilson 1987; O'Donnell and Dilks 1994; Matthews 1980; Moorhouse 1995, 53–54; Jackson 1960.

8. Oliver 1955, 548.

9. The warmest month on Kapiti Island is February, with a normal mean temperature of 17°C; the coldest month is July, at about 9°C (New Zealand Department of Lands and Survey 1981, 38).

10. IUCN/SSC 1993, 12–13, 15; Baillie and Groombridge 1996, 41.

11. N.Z. Dept. of Lands and Survey 1981, 59.

12. N.Z. Dept. of Lands and Survey 1981, 43–44; Veitch and Bell 1990.

13. Hill and Hill 1987, 302.

14. Beggs 1988; Beggs and Wilson 1987, 1991; Moorhouse 1995; Moorhouse and Greene 1995; O'Donnell and Dilks 1986, 1989, 1994.

15. Bond and Diamond 1992.

16. Moorhouse and Greene 1995.
17. Jackson 1963b, 175; Moorhouse 1992.
18. Moorhouse 1992.
19. Kaka vocalizations are described in Moorhouse 1992.
20. Jackson 1963b.
21. Moorhouse 1995, 37–62; Brejaart 1988, 22, 34.
22. Moorhouse 1995, 1–9; Jackson 1963a; McCaskill 1954.
23. The figures for keas are from Jackson 1963a. Norway rats destroy roughly a quarter of all kaka nestlings (Moorhouse 1995, 5).
24. Population densities for kakas are given in Moorhouse 1995, 2; IUCN/SSC 1993, 15. For keas, see Jackson 1960; Bond and Diamond 1992.
25. Moorhouse 1995, 55. Even fledgling kakas have a high survival rate. Of nine kakas that Moorhouse observed before and after fledging on Kapiti, all but three survived their first year. Two of the three had been weak and unhealthy at hatching and apparently succumbed immediately after fledging (Moorhouse, personal communication, 1990).
26. The birds' internal temperatures are given in McNab and Salisbury 1995. For the temperature regimes in Arthur's Pass and Kapiti, see Arthur's Pass National Park 1986, 26–27; N.Z. Dept. of Lands and Survey 1981, 38. Caloric requirements were calculated from a basal metabolic rate of 1 cc O_2/gh (McNab and Salisbury, 1995), assuming a caloric equivalent of 4.8 kcal/l O_2 (Gordon 1972, 51–52).
27. Jackson 1960, 1969; Bond and Diamond 1992.
28. Bond and Diamond 1992; Jackson 1963a; IUCN/SSC 1993, 20.
29. Jackson 1963a.
30. Jackson 1962a, 1963a.
31. Jackson 1969, 44. Nest visitation by adults other than the resident pair was noted by both Jackson (1963a) and Wilson (1990).
32. IUCN/SSC 1993, 37; Moorhouse and Greene 1995. For a general model of life history strategies in an unpredictable environment, see Horn and Rubenstein 1984. Delayed reproduction in birds is a common adaptation to a limited food supply; the classic reference is Lack

1968, 295–305. Similarly, Ashmole and Tovar (1968) infer that birds whose methods of obtaining food require great skill must lay small clutches, and only one clutch per year, as well as feed their young for a long period after they fledge.

33. Jackson 1960, 1969; Webb and Kelly 1993.

34. Moorhouse 1995, 37–62; Beggs and Wilson 1991; O'Donnell and Dilks 1986.

35. The relationship between the distribution of resources and the requirement for learning is generally accepted; see for example, Kamil and Yoerg 1982. In one of the few studies that compare the incidence of play between two closely related populations in contrasting environments, Berger (1979) found that bighorn sheep living in the desert play less than those living in more northern, mountainous regions. He attributed this difference not to the lack of resources in the desert but to the low population density, which made for smaller herd sizes.

CHAPTER SIX

1. Pullar 1991; Metcalfe 1990; "Angell Imprisoned" 1990.
2. "Acting Young" 1991.
3. Dixon 1986, 2; Low 1984, 11; Roet and Milliken 1985, 1–4; Beissinger and Snyder 1992, 222–30.
4. Salmon 1975.
5. IUCN/SSC 1993, 5; J. M. Cunningham 1948; R. Anderson 1986.
6. R. Anderson 1986; Heather and Robertson 1997, 356.
7. Low 1984, 144–48.
8. Dixon 1986, 2.
9. Stephen Phillipson, personal communication, 1991.
10. Ibid.; Aspinall 1990, 19–20.
11. Andrew Grant, Stephen Phillipson, personal communication, 1991.
12. IUCN/SSC 1993, 6.
13. Aspinall 1990, 19. See also Aspinall 1967.

14. Temple 1994.

15. See, for example, Strimback 1989; Temple 1994; idem 1996, 88–89.

16. Wilson 1990.

17. Soper 1965, 50; Arthur's Pass National Park 1986, 42; Peter Simpson, personal communication, 1988.

18. Peter Simpson, personal communication, 1990.

19. Peter Simpson, personal communication, 1991. See also Temple 1994; idem 1996, 72.

20. IUCN/SSC 1993, 10–11.

21. Zoos can be equally dangerous. Captive keas often suffer from lead poisoning, most likely from eating paint (Gray 1972).

22. N.Z. Dept. of Conservation 1989, 151–52.

23. King 1984, 154. For some time possums were protected by the Animals and Game Protection Act of 1921, a law that extended protection to a variety of introduced species in the interests of the fur industry. By 1930, however, possums had come to be regarded as pests, and they and most other introduced species lost their protected status (Salmon 1975, 653–59).

24. Bekoff 1978, 359–60; Mack 1981.

25. Spurr 1979; Atkinson and Cameron 1993.

26. Baillie and Groombridge 1996, 152, 181. On the kakapo, see Low 1984, 108–9; on the kaka, Beggs and Wilson 1991; Pierce, Atkinson, and Smith 1993; IUCN/SSC 1993, 12–13; on parakeets, Falla et al. 1978, 164–65; and on the Norfolk Island kaka, Greenway 1967, 312–13; Oliver 1930. The other parrot extinctions in modern times were the black-fronted parakeet from Tahiti (1844), the Cuban macaw from Cuba (1850s), Newton's parakeet from Rodrigues in the Mascarene Islands (1875), the Seychelles parakeet from Mahe and Silhouette Islands (1906), the Carolina conure from the southeastern United States (1926), and the glaucous macaw from South America (last date not recorded) (Low 1984, 154).

27. Wilson 1990; Wilson and Brejaart 1992.

28. Bull et al. 1985, 144, 146.

29. The Ornithological Society's database is a compilation of sightings rather than a count of individuals. R. Anderson (1986) noted that keas had been observed in 480 of the 1,927 quadrats on the South Island. Assuming that each quadrat could support a minimum of two keas, with greater numbers in some areas, he inferred a total kea population of between 1,000 and 5,000. His estimate of two keas per quadrat was an educated guess. Jackson (1960) and Clarke (1970) have obtained direct quantitative estimates of kea population densities in the South Island backcountry—about 0.0036 keas per hectare. If this figure is assumed to hold, on the average, for all of the 480 quadrats in which keas were observed, the maximum feasible wild population would be about 15,000 birds (Bond and Diamond 1992).

For a more precise estimate of density, we went back to the Ornithological Society's database and transcribed, for each quadrat, both the number of visits in which a kea was sighted and the number in which no keas were seen. Given these values, we calculated for each quadrat the probability of not sighting keas. If we assume that the number of kea sightings per visit is a Poisson variate, we can compute from the probability of not sighting any keas in a given quadrat, p, the expected (or mean) number of sightings per visit: $E(p) = e^{-p}$.

The local population density is a nonlinear function of the expected number of sightings, apparently because the number of birds in a sighting increases with population density; i.e., one sighting may consist of a single kea in a low-population area or a flock of ten in a region of high population density. To convert the expected number of sightings per visit into a true measure of local population density, we determined the sighting probabilities in three areas for which quantitative estimates of kea population density had previously been obtained and derived a numerical function that relates sightings per visit to the number of keas in the quadrat.

Over the entire South Island, the Ornithological Society database lists a total of 8,611 visits to 1,927 quadrats. Of these visits, 961 included kea sightings. Thus the expected number of sightings per visit for the island as a whole is 0.118. Our inferred relationship between sightings per visit and density yields an estimated mean of 1.59 keas per quadrat. If we multiply this figure by the 1,927 quadrats on the island, we obtain an estimate of 3,070 keas in the wild population. This value is gratifyingly close to other, independent estimates (Wilson 1990; R. Anderson 1986).

30. Wilson 1990; Marriner 1908, 135–41.

31. Marriner 1908, 41; J. M. Cunningham 1948.

32. Campbell 1976, 59, 61; Jackson 1960; Bond and Diamond 1992.

33. IUCN/SSC 1993, 31–32. Baillie and Groombridge (1996, 152) classify keas as "near threatened."

34. Temple (1994) quotes several sheep ranchers who agree that the kea's harassment of sheep is mainly motivated by play.

35. Mayr 1974.

36. Fagen 1982.

37. For evidence of imitation and observational learning in parrots, see Moore 1996; Galef, Manzig, and Field 1986; B. V. Dawson and Foss 1965. On the concept of cultural transmission, see Huffman 1996; Terkel 1996; Galef 1990; Cavalli-Sforza, Feldman, Chen, and Dornbusch 1982.

38. Many features of the kea's ecology, development, and social behavior may be echoed among the lower primates. Fragaszy and Visalberghi (1996), in a series of elegant laboratory investigations, have discovered that like keas, capuchin monkeys have a broad, generalist diet and a strong tendency to social facilitation, particularly among the young. Capuchins also show great tolerance toward young animals attempting to share their food. Again like keas, the adults evidently do not teach their offspring, and the young monkeys do not appear to learn their foraging techniques—which are not particularly complex—from their parents. Thus, the narrative we have developed concerning the evolution of flexibility in the kea may have a much wider applicability.

REFERENCES

Acland, L. G. D. 1975. *The early Canterbury runs.* 4th ed. Christchurch: Whitcoulls.

"Acting young" to win kea social standing. 1991. *National Geographic Magazine* 180(3): 4.

Allen, R. R., and K. H. Platt. 1990. Annual seedfall variation in *Nothofagus solandri* (Fagacea), Canterbury, New Zealand. *Access* 57: 199–206.

Anderson, A. 1982. Habitat preferences of moa in central Otago, A.D. 1000–1500, according to palaeobotanical and archaeological evidence. *Journal of the Royal Society of New Zealand* 12(3): 321–36.

———. 1984. The extinction of moa in southern New Zealand. In P. S. Martin and R. G. Klein, eds., *Quaternary extinctions: A prehistoric revolution,* 728–40. Tucson: University of Arizona Press.

———. 1989. *Prodigious birds.* Cambridge: Cambridge University Press.

Anderson, A., R. McGovern-Wilson, and S. Holdaway. 1991. Identification and analysis of faunal remains. In A. Anderson and R. McGovern-Wilson, eds., *Beech forest hunters,* 56–66. New Zealand

Archaeological Association Monograph no. 18. Auckland: New Zealand Archaeological Association.

Anderson, R. 1986. Keas for keeps. *Forest and Bird* 17(1): 2–5.

Andrews, J. R. H. 1986. *The southern ark: Zoological discovery in New Zealand, 1769–1900.* Honolulu: University of Hawaii Press.

Angell imprisoned for taking keas. 1990. *Canterbury Press.* Sept. 1, p. 6.

Archer, J. 1988. *The behavioural biology of aggression.* Cambridge: Cambridge University Press.

Arthur's Pass National Park. 1986. *The story of Arthur's Pass National Park.* Auckland: Arthur's Pass National Park.

Arthur's Pass National Park Board. 1958. *Handbook to the Arthur's Pass National Park.* Christchurch: Arthur's Pass National Park Board.

Ashmole, N. P., and S. H. Tovar. 1968. Prolonged parental care in royal terns and other birds. *Auk* 85: 90–100.

Aspinall, J. C. 1967. Some observations on keas. *New Zealand Tussock and Grassland Revue* 12: 14–17.

———. 1990. *Keas.* Wanaka, N.Z.: Author.

Atkinson, I. A. E., and E. K. Cameron. 1993. Human influence on the terrestrial biota and biotic communities of New Zealand. *Trends in Ecology and Evolution* 8(12): 447–51.

Baggaley, E. J. 1967. *A geography of New Zealand.* Melbourne: Thomas Nelson (Australia).

Baillie, J., and B. Groombridge. 1996. *The 1996 IUCN red list of threatened animals.* Gland, Switzerland: International Union for Conservation of Nature and Natural Resources.

Balda, R. P., and J. H. Balda. 1978. The care of young piñon jays (*Gymnorhinus cyanocephalus*) and their integration into the flock. *Journal für Ornithologie* 119: 146–71.

Barkan, C. P. L., J. L. Craig, S. D. Strahl, A. M. Stewart, and J. L. Brown. 1986. Social dominance in communal Mexican jays *Aphelocoma ultramarina. Animal Behaviour* 34: 175–87.

Bartholomew, G. A., Jr., R. C. Lasiewski, and E. C. Crawford Jr. 1968.

Patterns of panting and gular flutter in cormorants, pelicans, owls, and doves. *Condor* 70: 31–34.

Beattie, J. H. 1994. *Traditional lifeways of the southern Maori.* Dunedin, N.Z.: University of Otago Press.

Beauchamp, A. J., and T. H. Worthy. 1988. Forum on the decline and conservation of takahe. *Journal of the Royal Society of New Zealand* 18(1): 103–18.

Beggs, J. R. 1988. Energetics of kaka in a South Island beech forest. M.S. thesis, University of Auckland.

Beggs, J. R., and P. R. Wilson. 1987. Energetics of South Island kaka (*Nestor meridionalis*) feeding on the larvae of kanuka longhorn beetles (*Ochrocydus huttoni*). *New Zealand Journal of Ecology* 10: 143–47.

———. 1991. The kaka *Nestor meridionalis,* a New Zealand parrot endangered by introduced wasps and mammals. *Biological Conservation* 56: 23–38.

Beissinger, S. R., and N. F. R. Snyder. 1992. *New world parrots in crisis: Solutions from conservation biology.* Washington, D.C.: Smithsonian Institution Press.

Bekoff, M. 1972. The development of social interaction, play, and metacommunication in mammals: An ethological perspective. *Quarterly Review of Biology* 47(4): 412–34.

———. 1976. Animal play: Problems and perspectives. In P. P. G. Bateson and P. H. Klopfer, eds., *Perspectives in ethology,* 2: 165–88. New York: Plenum.

———. 1978. *Coyotes: Biology, behavior, and management.* New York: Academic Press.

———. 1984. Social play behavior. *Bioscience* 34(4): 228–33.

———. 1989. Behavioral development of terrestrial carnivores. In J. L. Gittleman, ed., *Carnivore behavior, ecology and evolution,* 89–124. Ithaca: Cornell University Press.

Belcham, A. 1988. An ecological study of the kea *(Nestor notabilis)* in the Southern Alps of New Zealand. M.S. thesis, Southampton University (England).

Benham, W. B. 1906. On the flesh-eating propensity of the kea. *Transactions of the New Zealand Institute* 39: 71–89.

Berger, J. 1979. Social ontogeny and behavioural diversity: Consequences for bighorn sheep *Ovis canadensis* inhabiting desert and mountain environments. *Journal of Zoology (London)* 188: 251–66.

Bergquist, C. A. L. 1985. Difference in the diet of male and female tui (*Prosthemadera novaeseelandiae:* Meliphagidae). *New Zealand Journal of Ecology* 12: 573–76.

Best, E. 1977. *Forest lore of the Maori.* Wellington: E. C. Keating.

Best, H., and R. Powlesland. 1985. *Kakapo.* Dunedin, N.Z.: John McIndoe.

Binney, D. H. 1990. Bird art in New Zealand. In B. J. Gill and B. D. Heather, eds., *A flying start: Commemorating fifty years of the Ornithological Society of New Zealand,* 107–9. Auckland: Random Century.

Bond, A. B., and J. Diamond. 1992. Population estimates of kea in Arthur's Pass National Park. *Notornis* 39: 151–60.

———. 1995. Sex differences in the kea, *Nestor notabilis,* using overlapping clustering. Paper read at 31st annual meeting of the Animal Behavior Society, 9–13 July, University of Nebraska, Lincoln.

Bond, A. B., K.-J. Wilson, and J. Diamond. 1991. Sexual dimorphism in the kea (*Nestor notabilis*). *Emu* 91: 12–19.

Brejaart, R. 1988. Diet and feeding behaviour of the kea (*Nestor notabilis*). M.S. thesis, Lincoln University, Canterbury (N.Z.).

Brewster, B. 1987. *Te moa: The life and death of New Zealand's unique bird.* Nelson, N.Z.: Nikau Press.

Brockman, H. J., and C. J. Barnard. 1976. Kleptoparasitism in birds. *Animal Behaviour* 27: 487–514.

Brown, J. L. 1963. Aggressiveness, dominance, and social organization in the steller jay. *Condor* 65: 460–84.

Brown, J. L., and E. R. Brown. 1984. Parental facilitation: Parent-offspring relations in communally breeding birds. *Behavioral Ecology and Sociobiology* 14: 203–9.

Bull, P. C., and A. H. Whitaker. 1975. The amphibians, reptiles, birds and mammals. In G. Kuschel, ed., *Biogeography and ecology in New Zealand*, 231–76. The Hague: Dr. W. Junk b.v.

Bull, P. C., P. D. Gaze, and C. J. R. Robertson. 1985. *The atlas of bird distribution in New Zealand*. Wellington North: The Ornithological Society of New Zealand.

Buller, W. L. 1869. Notes on the ornithology of New Zealand. *Transactions and Proceedings of the New Zealand Institute* 2: 385–92.

———. 1870. Further notes on the ornithology of New Zealand. *Transactions and Proceedings of the New Zealand Institute* 3: 37–56.

———. 1883. On some rare species of New Zealand birds. *Transactions of the New Zealand Institute* 16: 308–18.

———. 1894. Illustrations of Darwinism; or, the avifauna of New Zealand considered in relation to the fundamental law of descent with modification. *Transactions of the New Zealand Institute* 27: 75–104.

Burger, J. 1986. Effects of age on foraging in birds. In H. Ovellet, ed., *Acta XIX Congressus Internationalis Ornithologici*, 1:1127–40. Ottawa: University of Ottawa Press.

Burger, J., and M. Gochfield. 1979. Age differences in ring-billed gull kleptoparasitism on starlings. *Auk* 96: 806–8.

———. 1981. Age-related differences in piracy behaviour of four species of gulls, *Larus. Behaviour* 77: 242–67.

Burrows, C. J. 1980. Some empirical information concerning the diet of moas. *New Zealand Journal of Ecology* 3: 125–30.

Burton, P. J. K. 1974. Anatomy of head and neck in the huia *(Heteralocha acutirostris)* with comparative notes on other Callaeidae. *Bulletin of the British Museum (Natural History), Zoology* 27: 1–48.

Campbell, B. A. 1976. Feeding habits of the kea in the Routeburn Basin. M.S. thesis, University of Otago, Dunedin.

Cassels, R. 1984. The role of prehistoric man in the faunal extinctions of New Zealand and other Pacific islands. In P. S. Martin and R. G. Klein, eds., *Quaternary extinctions: A prehistoric revolution*, 741–67. Tucson: University of Arizona Press.

Caughley, G. 1988. The colonization of New Zealand by the Polynesians. *Journal of the Royal Society of New Zealand* 18(3): 245–70.

Cavalli-Sforza, L. L., N. W. Feldman, K. H. Chen, and S. M. Dornbusch. 1982. Theory and observation in cultural transmission. *Science* 218: 19–27.

Cheney, D. L., and R. M. Seyfarth. 1990. *How monkeys see the world.* Chicago: University of Chicago Press.

Child, P. 1975. Observations on altitudes reached by some birds in central and northwest Otago. *Notornis* 22: 143–50.

Christidis, L., R. Schodde, D. D. Shaw, and S. F. Maynes. 1991. Relationships among the Australo-papuan parrots, lorikeets, and cockatoos (Aves: Psittaciformes): Protein evidence. *Condor* 93: 302–17.

Clark, A. H. 1970. *The invasion of New Zealand by people, plants and animals.* Westport, Conn.: Greenwood Press.

Clarke, C. M. H. 1970. Observations on population, movements and food of the kea (*Nestor notabilis*). *Notornis* 17: 105–14.

Clayton, D. A. 1978. Socially facilitated behavior. *Quarterly Review of Biology* 53(4): 373–92.

Cooper, A., I. A. E. Atkinson, W. G. Lee, and T. H. Worthy. 1993. Evolution of the moa and their effect on the New Zealand flora. *Trends in Ecology and Evolution* 8(12): 433–37.

Cooper, A., C. Mourer-Chauviré, G. K. Chambers, A. von Haeseler, A. C. Wilson, and S. Pääbo. 1992. Independent origins of New Zealand moas and kiwis. *Proceedings of the National Academy of Sciences (U.S.)* 89: 8741–44.

Cooper, R. A., and P. R. Millener. 1993. The New Zealand biota: Historical background and new research. *Trends in Ecology and Evolution* 8(12): 429–33.

Cowen, J. 1905. Notes on some South Island birds, and maori associations connected therein. *Transactions of the New Zealand Institute* 38: 337–41.

Cracraft, J. 1980. Moas and the maori. *Natural History* 89(10): 28–36.

Craig, J. L. 1985. Status and foraging in New Zealand honeyeaters. *New Zealand Journal of Ecology* 12: 589–97.

Cunningham, A. 1974. Kea observations in the Tararua Range. *Notornis* 21: 382–83.

Cunningham, J. M. 1948. Number of keas. *Notornis* 2: 154.

Dadelszen, E. J. von. 1904. *The New Zealand official yearbook, 1904.* Wellington: John Mackay, Government Printer.

Darwin, C. 1859. *On the origin of species by means of natural selection; or, the preservation of favoured races in the struggle for life.* London: Murray.

Daugherty, C. H., G. W. Gibbs, and R. A. Hitchmough. 1993. Mega-island or micro-continent? New Zealand and its fauna. *Trends in Ecology and Evolution* 8(12): 437–42.

Davidson, J. M. 1984. *The prehistory of New Zealand.* Auckland: Longman Paul.

Dawson, B. V., and B. M. Foss. 1965. Observational learning in budgerigars. *Animal Behaviour* 13: 470–74.

Dawson, E. 1959. The supposed occurrence of kakapo, kaka and kea in the Chatham Islands. *Notornis* 8: 106–15.

Deckert, G., and K. Deckert. 1982. Spielverhalten und Komfortbewegungen beim Grünflügelara (*Ara chloroptera* G. R. Gray). *Bonner Zoologische Beiträge* 33(2–4): 269–81.

Derscheid, J. M. 1947. Strange parrots. I. The kea (*Nestor notabilis* Gould). *Avicultural Magazine* 53: 44–50.

Diamond, J., and A. B. Bond. 1989a. Curious kea. *Zoonooz* 61(12): 5–9.

———. 1989b. Note on the lasting responsiveness of a kea *Nestor notabilis* toward its mirror image. *Avicultural Magazine* 89(2): 92–94.

———. 1991. Social behavior and the ontogeny of foraging in the kea (*Nestor notabilis*). *Ethology* 88: 128–44.

Diamond, J. M. 1990a. Bob Dylan and moas' ghosts. *Natural History* 99(10): 26–31.

———. 1990b. New Zealand as an archipelago: An international perspective. In D. R. Towns, C. H. Daughrtey, and I. A. E. Atkinson, eds., *Ecological restoration of New Zealand islands,* 3–8. Conservation Sciences Publication no. 2. Wellington: Department of Conservation.

Diamond, J. M., and C. R. Veith. 1981. Extinctions and introductions in the New Zealand avifauna: Cause and effect? *Science* 211: 499–501.

Dixon, A. M. 1986. *Evaluation of the psittacine importation process in the United States.* Washington, D.C.: World Wildlife Fund.

Drury, W. H., and W. J. Smith. 1968. Defense of feeding areas by adult herring gulls and intrusion by young. *Evolution* 22: 193–201.

English, W. L. 1939. Keas and sheep. *Avicultural Magazine* 4: 63–64.

Fääborg, J., and J. C. Bednarz. 1990. Galápagos and Harris' hawks: Divergent causes of sociality in two raptors. In P. B. Stacey and W. D. Koenig, eds., *Cooperative breeding in birds: Long-term studies of ecology and behavior,* 359–83. Cambridge: Cambridge University Press.

Fagen, R. 1981. *Animal play behavior.* New York: Oxford University Press.

———. 1982. Evolutionary issues in development of behavioral flexibility. In P. P. G. Bateson and P. H. Klopfer, eds., *Perspectives in ethology,* 5: 365–84. New York: Plenum.

Falla, R. A., R. B. Sibson, and E. G. Turbott. 1978. *The new guide to the birds of New Zealand.* Auckland: Collins.

Feduccia, A. 1996. *The origin and evolution of birds.* New Haven: Yale University Press.

Ficken, M. S. 1977. Avian play. *Auk* 94: 573–82.

Fleming, C. A. 1962. History of the New Zealand land bird fauna. *Notornis* 9: 270–74.

———. 1975. The geological history of New Zealand and its biota. In G. Kuschel, ed., *Biogeography and ecology in New Zealand,* 1–86. The Hague: Dr. W. Junk b.v.

———. 1979. *The geological history of New Zealand and its life.* Auckland: Auckland University Press.

———. 1980. Orange-fronted parakeet: Record of flocking. *Notornis* 27: 388–407.

Forshaw, J. M. 1977. *Parrots of the world.* Neptune, N.J.: T.F.H. Publications.

Forster, R. R. 1975. The spiders and harvestmen. In G. Kuschel, ed., *Biogeography and ecology in New Zealand,* 493–505. The Hague: Dr. W. Junk b.v.

Fragaszy, D. M., and E. Visalberghi. 1996. Social learning in monkeys: Primate "primacy" reconsidered. In C. M. Heyes and B. G. Galef Jr., eds., *Social learning in animals: The roots of culture,* 65–84. San Diego: Academic Press.

Galef, B. G., Jr. 1988. Imitation in animals: History, definition, and interpretation of data from the psychological laboratory. In T. R. Zentall and B. G. Galef Jr., eds., *Social learning: Psychological and biological perspectives,* 3–28. Hillsdale, N.J.: Lawrence Erlbaum Associates.

———. 1990. Tradition in animals: Field observations and laboratory analysis. In M. Bekoff and D. Jamieson, eds., *Interpretation and explanation in the study of animal behavior,* 1: 74–95. Boulder, Colo.: Westview.

———. 1996. Introduction. In C. M. Heyes and B. G. Galef Jr., eds., *Social learning in animals: The roots of culture,* 3–15. San Diego: Academic Press.

Galef, B. G., Jr., L. A. Manzig, and R. M. Field. 1986. Imitation learning in budgerigars: Dawson and Foss (1965) revisited. *Behavioural Processes* 13: 191–202.

Garrod, A. H. 1872. Note on the tongue of the psittacine genus *Nestor. Proceedings of the Zoological Society of London* 40: 787–89.

Gaston, A. J. 1977. Social behavior with groups of jungle babblers (*Turdoides striatus*). *Animal Behaviour* 25: 828–48.

Gaze, P. D., and M. N. Clout. 1983. Honeydew and its importance to birds in beech forests of South Island, New Zealand. *New Zealand Journal of Ecology* 6: 33–37.

Geist, V. 1978. *Life strategies, human evolution, environmental design: Towards a biological theory of health.* New York: Springer-Verlag.

Gibb, J. A. 1990. Fifty years of ornithology in New Zealand. In B. J. Gill and B. D. Heather, *A flying start: Commemorating fifty years of the Ornithological Society of New Zealand,* 79–83. Auckland: Random Century.

Gibson, J. J. 1966. *The senses considered as perceptual systems.* Boston: Houghton Mifflin.

Godly, E. G. 1975. Flora and vegetation. In G. Kuschel, ed., *Biogeography and ecology in New Zealand,* 177–299. The Hague: Dr. W. Junk b.v.

Gordon, M. S. 1972. *Animal physiology.* New York: Macmillan.

Gould, J. 1840–48. *Birds of Australia.* 7 vols. London: Author.

———. 1856. On two new species of birds (*Nestor notabilis* and *Spatula variegata*) from the collection of Walter Mantell, Esq. *Proceedings of the Zoological Society of London* 24: 94–95.

Gray, C. 1972. Lead toxicosis. *Journal of Zoo Animal Medicine* 3(1): 20.

Green, R. C. 1975. Adaptation and change in Maori culture. In G. Kuschel, ed., *Biogeography and ecology in New Zealand,* 591–641. The Hague: Dr. W. Junk b.v.

Greenway, J. C. 1967. *Extinct and vanishing birds of the world.* New York: Dover.

Greenwood, R. M., and A. E. Atkinson. 1977. Evolution of divaricating plants in New Zealand in relation to moa browsing. *Proceedings of the New Zealand Ecological Society* 24: 21–33.

Guest, R. 1975. Forest-dwelling birds of the Wairau catchment, Marborough. *Notornis* 22: 23–26.

Gwinner, E. 1966. Untersuchungen über das Ausdrucks- and Sozialverhalten des Kolkraben (*Corvus corax* L.). *Zeitschrift für Tierpsychologie* 21: 657–748.

Heather, B., and H. Robertson. 1997. *The field guide to the birds of New Zealand.* Oxford: Oxford University Press.

Heinrich, B. 1989. *Ravens in winter.* New York: Summit Books.

Higuchi, H. 1986. Bait fishing by green-backed heron *Ardeola striata* in Japan. *Ibis* 128: 285–90.

Hill, S., and J. Hill. 1987. *Richard Henry of Resolution Island.* Dunedin, N.Z.: John McIndoe.

Hinde, R. A. 1974. *Biological bases of human social behaviour.* New York: McGraw-Hill.
———. 1981. Animal signals: Ethological and games-theory approaches are not incompatible. *Animal Behaviour* 29: 535–42.
Hockey, P. A. R., P. G. Ryan, and A. L. Bosman. 1989. Age-related intraspecific kleptoparasitism and foraging success of kelp gulls *Larus dominicanus*. *Ardea* 77: 205–10.
Holdaway, R. N. 1989a. New Zealand's pre-human avifauna and its vulnerability. *New Zealand Journal of Ecology*, suppl. 12: 11–25.
———. 1989b. Terror of the forests. *New Zealand Geographic* 4: 56–63.
Holdaway, R. N., and T. H. Worthy. 1993. First North Island fossil record of kea, and morphological and morphometric comparison of kea and kaka. *Notornis* 40: 95–108.
Holdaway, S. 1991. Bird-skin and feathers. In A. Anderson and R. McGovern-Wilson, eds., *Beech Forest Hunters*, 67–71. New Zealand Archaeological Association Monograph no. 18. Auckland: New Zealand Archaeological Association.
Holyoak, D. T. 1973. Comments on the taxonomy and relationship in the parrot subfamilies Nestorinae, Loriinae, and Platycerinae. *Emu* 73: 157–76.
Horn, H. S., and D. I. Rubenstein. 1984. Behavioural adaptations and life history. In J. R. Krebs and N. B. Davis, eds., *Behavioural ecology: An evolutionary approach*, 270–98. Sunderland, Mass.: Sinaur Associates.
Huffman, M. A. 1996. Acquisition of innovative cultural behaviors in nonhuman primates: A case study of stone handling, a socially transmitted behavior in Japanese macaques. In C. M. Heyes and B. G. Galef Jr., eds., *Social learning in animals: The roots of culture*, 267–89. San Diego: Academic Press.
Huntingford, F. A., and A. K. Turner. 1987. *Animal conflict.* London: Chapman and Hall.
Huxley, V. J. 1963. Lorenzian ethology. *Zeitschrift für Tierpsychologie* 20: 401–9.
IUCN/SSC (International Union for the Conservation of Natural Resources, Species Survival Commission) Captive Breeding Specialist

Group. 1993. *Kea-kaka population viability assessment*. Christchurch: New Zealand Department of Conservation.

Jackson, J. R. 1960. Keas at Arthur's Pass. *Notornis* 9: 39–59.

———. 1962a. Do keas attack sheep? *Notornis* 10: 33–38.

———. 1962b. Life of the kea. *Canterbury Mountaineer* 31: 120–23.

———. 1963a. The nesting of keas. *Notornis* 10: 319–26.

———. 1963b. Studies at a kaka's nest. *Notornis* 10: 168–75.

———. 1969. What do keas die of? *Notornis* 16: 33–44.

Johnson, S. J. 1986. Development of hunting and self-sufficiency in juvenile red-tailed hawks (*Buteo jamaicensis*). *Raptor Research* 20: 29–34.

Kamil, A. C., and S. I. Yoerg. 1982. Learning and foraging behavior. In P. P. G. Bateson and P. H. Klopfer, eds., *Perspectives in ethology*, 5: 325–64. New York: Plenum.

Kavanau, J. L. 1987. *Lovebirds, cockatiels, budgerigars: Behavior and evolution*. Los Angeles: Science Software Systems.

Keller, R. 1972. A few observations on a kea-family *Nestor notabilis* during a short stay at the Jersey Wildlife Preservation Trust. In *Jersey Wildlife Preservation Trust Ninth Annual Report*, 54–56. Jersey: Jersey Wildlife Preservation Trust.

———. 1975. Das Spielverhalten der Keas (*Nestor notabilis* Gould) des Zürcher Zoos. *Zeitschrift für Tierpsychologie* 38: 393–408.

———. 1976. Beitrag zur Biologie and Ethologie der Keas (*Nestor notabilis* Gould) des Zürcher Zoos. *Zoologische Beiträge* 22(1): 111–56.

Kikkawa, J. 1966. Population distribution of land birds in temperate rainforest in southern New Zealand. *Transactions of the Royal Society of New Zealand* 7(17): 215–77.

Kilham, L. 1989. *The American crow and the common raven*. College Station: Texas A&M University Press.

King, C. 1984. *Immigrant killers*. Auckland: Oxford University Press.

Kirk, E. J., R. G. Powlesland, and S. C. Cork. 1993. Anatomy of the mandibles, tongue, and alimentary tract of kakapo, with some comparative information from kea and kaka. *Notornis* 40: 45–63.

Kirk, T. 1895. The displacement of species in New Zealand. *Transactions of the New Zealand Institute* 28: 1–27.

Kubat, S. 1992. Die Rolle von Neuigkeit, Andersartigkeit, und sozialer Struktur für die Exploration von Objekten beim Kea *(Nestor notabilis)*. Ph.D. diss., University of Vienna.

Lack, D. 1968. *Ecological adaptations for breeding in birds.* London: Methuen.

Lavers, R., and J. Mills. 1984. *Takahe.* Dunedin, N.Z.: John McIndoe.

Leach, H. M. 1969. *Subsistence patterns in prehistoric New Zealand.* Studies in prehistoric anthropology, no. 2. Dunedin, N.Z.: Anthropology Department, University of Otago.

Levy, E. B. 1970. *Grasslands of New Zealand.* Wellington: A. R. Shearer.

Lint, K. C. 1958. High haunts and strange habits of the kea. *Zoonooz* 31(9): 3–6.

Lockie, J. D. 1956. Winter fighting in feeding flocks of rooks, jackdaws and carrion crows. *Bird Study* 3: 180–90.

Lorenz, K. 1976. Psychology and phylogeny. In J. Bruner, A. Jolly, and K. Sylva, eds., *Play: Its role in development and evolution,* 84–95. New York: Basic Books.

Loveland, K. A. 1986. Discovering the affordances of a reflecting surface. *Developmental Review* 6: 1–24.

Low, R. 1984. *Endangered parrots.* Poole, England: Blandford Press.

Mack, S. 1981. Battle renewed over coyote poison. *Science* 213: 849.

MacLean, A. A. E. 1986. Age-specific foraging ability and the evolution of deferred breeding in three species of gulls. *Wilson Bulletin* 98(2): 267–79.

Mallet, M. 1973. Nesting of the kea, *Nestor notabilis,* at Jersey Zoo. *Avicultural Magazine* 79(4): 122–26.

Marriner, G. B. 1906. Notes on the natural history of the kea with special references to its reputed sheep killing propensity. *Transactions of the New Zealand Institute* 39: 271–306.

———. 1907. Additional notes on the kea. *Transactions of the New Zealand Institute* 40: 534–37.

———. 1908. *The kea: A New Zealand problem.* Christchurch: Marriner Bros.

Martin, P., and T. M. Caro. 1985. On the functions of play and its role in behavioral development. In J. S. Rosenblatt, C. Beer, M. C. Busnel, and P. J. B. Slater, eds., *Advances in the study of behavior,* 15: 59–103. New York: Academic Press.

Matthews, P. J. 1980. Patterns of feeding on *Dysoxylum* and *Planchonella* fruits. *Tane* 26: 79–82.

Mayr, E. 1966. *Animal species and evolution.* Cambridge: Harvard University Press.

———. 1974. Behavior programs and evolutionary strategies. *American Scientist* 62: 650–59.

McCann, C. 1963. External features of the tongues of New Zealand Psittaciformes. *Notornis* 10: 326–45.

McCaskill, J. R. 1954. A kea's nest. *Notornis* 6: 24–25.

McEwen, W. M. 1978. The food of the New Zealand pigeon (*Hemiphaga novaeseelandiae novaeseelandiae*). *New Zealand Journal of Ecology* 1: 99–108.

McGlone, M. S. 1988. New Zealand. In B. Huntley and T. Webb III, eds., *Vegetation history,* 557–99. Dordrecht: Kluwer Academic.

———. 1989. The Polynesian settlement of New Zealand in relation to environmental and biotic changes. *New Zealand Journal of Ecology,* suppl. 12: 115–29.

McKay, A. 1884. On the kea or mountain parrot. *Transactions of the New Zealand Institute* 17: 449.

McMillan, P., A. Hart, D. Robertson, A. Stewart, and R. Webber. 1989. Creatures of the deep ocean. *New Zealand Geographic* 4: insert.

McNab, B. K., and C. A. Salisbury. 1995. Energetics of New Zealand's temperate parrots. *New Zealand Journal of Zoology* 22: 339–49.

Meads, M. J., K. J. Walker, and G. P. Elliott. 1984. Status, conservation, and management of the land snails of the genus *Powelliphanta* (Mollusca: Pulmonata). *New Zealand Journal of Zoology* 11: 277–306.

Medway, D. 1976. Extant types of New Zealand birds from Cook's voyages. *Notornis* 23: 120–37.

Menzies, [Hon. Dr.]. 1878. Memorandum of the kea. *Transactions of the New Zealand Institute* 11: 376–77.

Metcalfe, T. 1990. Rescued keas settling in new environment. *Canterbury Press.* Sept. 28, p. 2.

Millener, P. R. 1988. Contributions to New Zealand's late Quaternary avifauna, I: *Pachyplichas,* a new genus of wren (Aves: Acanthisittidae), with two new species. *Journal of the Royal Society of New Zealand* 18(4): 383–406.

———. 1990. Evolution, extinction and the subfossil record of New Zealand's avifauna. In B. J. Gill and B. D. Heather, *A flying start: Commemorating fifty years of the Ornithological Society of New Zealand,* 93–100. Auckland: Random Century.

Mitchell, J. 1981. Keas: call them irresistible. *Zoonooz* 54(5): 10–12.

Mitchell, R. W. 1991. Bateson's concept of metacommunication in play. *New Ideas in Psychology* 9(1): 73–87.

Mitchell, R. W., and N. S. Thompson. 1986. Deception in play between dogs and people. In R. W. Mitchell and N. S. Thompson, eds., *Deception: Perspectives on human and nonhuman deceit,* 193–204. Albany: State University of New York Press.

Moore, B. R. 1996. The evolution of imitative learning. In C. M. Heyes and B. G. Galef Jr., eds., *Social learning in animals: The roots of culture,* 245–65. San Diego: Academic Press.

Moorhouse, R. J. 1992. Der Kaka: Der "gewöhnliche" Papagei in Neuseelands Wäldern. *Papageien* 5: 90–95, 122–26.

———. 1995. Productivity, sexual dimorphism and diet of North Island kaka *(Nestor meridionalis septentrionalis)* on Kapiti Island. Ph.D. diss., Victoria University, Wellington.

Moorhouse, R. J., and T. C. Greene. 1995. Identification of fledgling and juvenile kaka *(Nestor meridionalis). Notornis* 42: 187–202.

Moorhouse, R. J., M. J. Sibley, B. D. Lloyd, and T. C. Greene. Forthcoming. Sexual dimorphism in the North Island kaka (*Nestor meridionalis septentrionalis*). *Ibis*.

Moreau, R. E., and W. M. Moreau. 1944. Do young birds play? *Ibis* 86: 93–94.

Morris, R., and H. Smith. 1988. *Wild south: Saving New Zealand's endangered birds*. Auckland: TVNZ.

Müller-Schwarze, D. 1978. *Evolution of play behavior*. Stroudsburg, Pa.: Dowden, Hutchinson and Ross.

Myers, J. G. 1924. The relation of birds to agriculture in New Zealand, X: The kea or mountain parrot. *New Zealand Journal of Agriculture* 29: 25.

New Zealand Department of Conservation. 1989. *Arthur's Pass National Park: Draft management plan*. Christchurch: National Parks and Reserves Board.

New Zealand Department of Lands and Survey. 1981. *Kapiti Island nature reserve management plan*. Department of Lands and Survey, Management Plan Series no. NR1. Wellington: Department of Lands and Survey.

O'Brien, C. 1981. *A book of New Zealand wildlife*. Auckland: Landsdowne Press.

O'Donnell, C. F. J., and P. J. Dilks. 1986. *Forest birds in South Westland: Status, distribution and habitat use*. New Zealand Wildlife Service, Department of Internal Affairs, Occasional Publication no. 10. Wellington: New Zealand Wildlife Service.

———. 1989. Sap-feeding by the kaka (*Nestor meridionalis*) in South Westland, New Zealand. *Notornis* 36: 64–71.

———. 1994. Foods and foraging of forest birds in temperate rain forest, South Westland, New Zealand. *New Zealand Journal of Ecology* 18: 87–107.

Oliver, W. R. B. 1930. *New Zealand birds*. Wellington: Fine Arts (N.Z.).

———. 1955. *New Zealand birds*. 2d ed. Wellington: A. H. and A. W. Reed.

Orbell, M. 1985. *The natural world of the Maori*. Auckland: David Bateman.

Ornithological Society of New Zealand. 1970. *Annotated checklist of the birds of New Zealand*. Wellington: A. H. and A. W. Reed.

Ortega, J. C., and M. Bekoff. 1987. Avian play: Comparative evolutionary and developmental trends. *Auk* 104: 338–41.

Pellis, S. M. 1981. A description of social play by the Australian magpie *Gymnorhina tibicen* based on Eshkol-Wachman notation. *Bird Behaviour* 3: 61–79.

———. 1982. Development of head and foot coordination in the Australian magpie *Gymnorhina tibicen*, and the function of play. *Bird Behaviour* 4: 57–62.

Pettigrew, J. B. 1908. *Design in nature*. Vol. 3. London: Longmans, Green.

Phillipps, W. J. 1963. *The book of the huia*. Christchurch: Whitcombe and Tombs.

———. 1966. *Maori life and custom*. Wellington: A. H. Reed and A. W. Reed.

Pierce, R. J., R. Atkinson, and E. Smith. 1993. Changes in bird numbers in six Northland forests, 1979–1993. *Notornis* 40: 285–93.

Porter, R. E. R., and D. G. Dawson. 1968. Reactions of birds to falcons. *Notornis* 15: 237.

Porter, S. 1947. The breeding of the kea, *Nestor notabilis*. *Avicultural Magazine* 53: 50–55.

Potts, K. J. 1969. Ethological studies of the kea (*Nestor notabilis*) in captivity: Nonreproductive behavior. B.S. thesis., Victoria University, Wellington.

———. 1976. Comfort movements of the kea, *Nestor notabilis* (Psittaciformes: Nestoridae). *Notornis* 23: 302–9.

———. 1977. Some observations of the agonistic behavior of the kea, *Nestor notabilis* (Nestoridae) in captivity. *Notornis* 24: 31–40.

Potts, T. H. 1870. On the birds of New Zealand. *Transactions of the New Zealand Institute* 3: 59–109.

———. 1871. The kea: Progress of development. *Nature* 4: 489.

———. 1976. *Out in the open: A budget of scraps of natural history gathered in New Zealand* [1882]. Christchurch: Capper Press.

Pullar, T. 1991. The kea (*Nestor notabilis*) in captivity. *Thylacinus* 16(2): 7–9.

Reed, A. H., and A. W. Reed. 1969. *Captain Cook in New Zealand*. Wellington: A. H. and A. W. Reed.

Reischek, A. 1885. Observations on the habits of some New Zealand birds, their usefulness or destructiveness to the country. *Transactions of the New Zealand Institute* 18: 96–104.

Riney, T., J. S. Watson, C. Bassett, E. G. Turbott, and W. E. Howard. 1959. *Lake Monk expedition: An ecological study in southern Fiordland*. New Zealand Department of Scientific and Industrial Research, Bulletin 135. Wellington: R. E. Owen.

Ritzmeier, M. 1995. The influence of hunger and low protein diet on exploration in Keas *(Nestor notabilis)*. M.S. thesis, University of Vienna.

Roet, E., and T. Milliken. 1985. *The Japanese psittacine trade*. Washington, D.C.: Traffic (Japan) and World Wildlife Fund (U.S.).

Rowell, T. E. 1966. Hierarchy in the organization of a captive baboon group. *Animal Behaviour* 14: 430–43.

Salmon, J. T. 1975. The influence of man on the biota. In G. Kuschel, ed., *Biogeography and ecology in New Zealand*, 643–61. The Hague: Dr. W. Junk b.v.

———. 1985. *Collins guide to the alpine plants of New Zealand*. Auckland: Wm. Collins.

Scarlett, R. J. 1979. Avifauna and man. In A. Anderson, ed., *Birds of a feather: Osteological and archaeological papers from the South Pacific in honour of R. J. Scarlett*, 75–90. New Zealand Archaeological Association Monograph no. 11. Oxford: BAR.

Schifter, H. 1965. Zucht von Keas im Züricher Zoo. *Freunde des Kölner Zoo* 8(2): 66–68.

Schmidt, C. R. 1971. Breeding keas *Nestor notabilis* at Zurich Zoo. *International Zoo Yearbook* 11: 137–40.

Sherry, D. F., and B. G. Galef Jr. 1990. Social learning without imitation: More about milk bottle opening by birds. *Animal Behaviour* 40: 987–89.

Shufeldt, R. W. 1918. The skeleton of the "kea parrot" of New Zealand (*Nestor notabilis*). *Emu* 18(1): 25–43.

Sibley, C. G., and J. E. Ahlquist. 1990. *Phylogeny and Classification of Birds.* New Haven: Yale University Press.

Sieber, J. 1983. Second generation captive breeding of the kea, *Nestor notabilis,* at Wilhelminenberg (Vienna, Austria). *Avicultural Magazine* 89(2): 71–73.

Skeate, S. T. 1985. Social play behaviour in captive white-fronted amazon parrots *Amazona albifrons. Bird Behaviour* 6: 46–48.

Skutch, A. F. 1976. *Parent birds and their young.* Austin: University of Texas Press.

Smith, G. A. 1971. The use of the foot in feeding, with especial reference to parrots. *Avicultural Magazine* 77: 93–100.

———. 1975. Systematics of parrots. *Ibis* 117: 18–68.

Smith, W. W. 1888. On the birds of the Lake Brunner district. *Transactions of the New Zealand Institute* 21: 205–24.

Snyder, N. F. R., J. W. Wiley, and C. B. Kepler. 1987. *The parrots of Luquillo: Natural history and conservation of the Puerto Rican parrot.* Los Angeles: Western Foundation of Vertebrate Zoology.

Soper, M. F. 1965. *More New Zealand bird portraits.* Christchurch: Whitcombe and Tombs.

Sprecht, R. L., M. E. Dettmann, and D. M. Jarzen. 1992. Community associations and structure in the Late Cretaceous vegetation of southeast Australasia and Antarctica. *Palaeogeography, Palaeoclimatology, Palaeoecology* 94: 283–309.

Spurr, E. B. 1979. A theoretical assessment of the ability of bird species to recover from an imposed reduction in numbers with particular reference to 1080 poisoning. *New Zealand Journal of Ecology* 2: 46–61.

Steadman, D. 1995. Prehistoric extinctions of Pacific island birds: Biodiversity meets zooarcheology. *Science* 267: 1123–31.

Stewart, K. 1984. *Collins handguide to the native trees of New Zealand.* Auckland: Wm. Collins.

Strimback, R. 1989. Invasion of the brekkie snatchers. *Living Bird* 8(1): 22–25.

Tebbich, S., M. Taborsky, and H. Winkler. 1996. Social manipulation causes cooperation in keas. *Animal Behaviour* 52: 1–10.

Temple, P. 1986. *The legend of the kea, or how Kritka stole the best beak and best claws from Ka, the great bird of all birds, and took the keas to live in the highest mountains.* Auckland: Hodder and Stoughton.

———. 1994. Kea: The feisty parrot. *New Zealand Geographic* 24: 102.

———. 1996. *Book of the kea.* Auckland: Hodder Moa Beckett.

Terkel, J. 1996. Cultural transmission of feeding behavior in the black rat (*Rattus rattus*). In C. M. Heyes and B. G. Galef Jr., eds., *Social learning in animals: The roots of culture,* 17–47. San Diego: Academic Press.

Tinbergen, N. 1960a. The evolution of behavior in gulls. *Scientific American* 203(6): 118–30.

———. 1960b. *The herring gull's world.* NewYork: Basic Books.

———. 1963. On aims and methods of ethology. *Zeitschrift für Tierpsychologie* 20: 410–33.

———. 1973. On appeasement signals [1959]. In N. Tinbergen, *The animal in its world,* 2: 113–29. Cambridge: Harvard University Press.

Travers, W. T. L. 1883. Some remarks on the distribution of the organic productions of New Zealand. *Transactions of the New Zealand Institute* 16: 461–67.

Trotter, M., and B. McCulloch. 1984. Moas, men and middens. In P. S. Martin and R. G. Klein, eds., *Quaternary extinctions: A prehistoric revolution,* 708–27. Tucson: University of Arizona Press.

———. 1989. *Unearthing New Zealand.* Wellington: Government Printing Office.

Turbott, E. G. 1967. *Buller's birds of New Zealand.* Honolulu: East-West Center Press.

Turner, E. R. A. 1965. Social feeding in birds. *Behaviour* 24: 1–46.

Van Dongen, M. W. M., and L. E. M. De Boer. 1984. Chromosome studies of eight species of parrots of the families Cacatuidae and Psittacidae (Aves: Psittaciformes). *Genetica* 65: 109–17.

Veitch, C. R., and B. D. Bell. 1990. Eradication of introduced animals from the islands of New Zealand. In D. R. Towns, C. H. Daugherty, and I. A. E. Atkinson, eds., *Ecological restoration of New Zealand islands*, 137–46. Conservation Sciences Publication no. 2. Wellington: New Zealand Department of Conservation.

Wallace, A. R. 1891. *Darwinism*. London: MacMillan.

Wallace, R. A. 1978. Social behavior on islands. In P. P. G. Bateson and P. H. Klopfer, eds., *Perspectives in ethology*, 3: 167–204. New York: Plenum.

Watters, R. F. 1965. *Land and society in New Zealand*. Wellington: A. H. and A. W. Reed.

Webb, C. J., and D. Kelly. 1993. The reproductive biology of the New Zealand flora. *Trends in Ecology and Evolution* 8(12): 442–47.

Westland National Park. 1982. *Fox Glacier Valley guide*. Franz Josef Glacier: Westland National Park.

White, T. 1894. The kea (*Nestor notabilis*), a sheep-eating parrot. *Transactions of the New Zealand Institute* 27: 273–80.

Wilson, K.-J. 1990. Kea: Creature of curiosity. *Forest and Bird* 21: 20–26.

Wilson, K.-J., and R. Brejaart. 1992. The kea: A brief research review. In L. Joseph, ed., *Issues in the conservation of parrots in Australasia and Oceania: Challenges to conservation biology*, 24–28. Royal Australasian Ornithologists Union Report no. 83. Moonee Ponds, Victoria: Royal Australasian Ornithologists Union.

Wodzicki, K., and S. Wright. 1984. Introduced birds and mammals in New Zealand and their effect on the environment. *Tuatara* 27(2): 77–104.

Worthy, T. H. 1989. Moas of the subalpine zone. *Notornis* 36: 191–96.

Worthy, T. H., and R. N. Holdaway. 1993. Quaternary fossil faunas from caves in the Punakaiki area, West Coast, South Island, New Zealand. *Journal of the Royal Society of New Zealand* 23(3): 147–254.

Worthy, T. H., and D. C. Mildenhall. 1989. A late Otiran-Holocene paleoenvironment construction based on cave excavations in northwest Nelson, New Zealand. *New Zealand Journal of Geology and Geophysics* 32: 243–53.

Yealland, J. 1941. Some parrot-like birds at Sterrebeek. *Avicultural Magazine* 5(12): 288–93.

Zajonc, R. B. 1965. Social facilitation. *Science* 149: 269–74.

Zeigler, H. P. 1975. Some observations on the development of feeding in captive kea (*Nestor notabilis*). *Notornis* 22: 131–34.

Zentall, T. R. 1996. An analysis of imitative learning in animals. In C. M. Heyes and B. G. Galef Jr., eds., *Social learning in animals: The roots of culture*, 221–43. San Diego: Academic Press.

INDEX

Italic page references denote illustrations and tables.

abundance. *See* population, kaka; population, kea
activity patterns, daily, 56–57, 64–65, 80–81; age and sex differences, 64, 80, *159*
affordances, 77–79, 98, 190n. 41
age determination: in kakas, 104–5; in keas, 51–52, 104–5
aggression: age and sex differences, 59, 80–81, 86, 93–94, 119, *160–61,* 187n. 17; behavior patterns, 58–60, 70–71, 86–88, 91–92; dominance maintained by, 59; in fledgling keas, 86–88; interspecific, 62; in kakas, 110–11, 114; during play, 73, 75, 77, 86, *172,* 189–90n. 33
allofeeding. *See* feeding, regurgitant

allopreening, 65–69, *65,* 76, 86; in kakas, 114
alpine habitat: changes in, 41, 179n. 29; description, 23, 48, 62; use by keas, 15–16, 24, 39, 42, 105, 112, 114–15, 117, 120–22, 135, 147, 150, 180n. 33. *See also* southern beech
Antarctica, 2, 19
appeasement, 91–92, 191n. 9. *See also* hunch display
Arthur's Pass National Park: climate, 48–49, 117, 185n. 6; fauna, 62, 116, 141, 143, 146; kea damage in, 135–38; physical description, 28, *29,* 46–48, 185n. 4; smuggling in, 123–25
Australia, 19–21, 31, 178n. 25, 184n. 44

bats, 20, 128, 178n. 23
Bealey River, *29,* 48–49

223

beech. *See* southern beech
Beggs, Jacqueline, 110
behavior, forms of. *See* ethogram
Belligny, Sainte Croix de, 30
bill. *See under* morphology
birds: extinct New Zealand species, 43–45; list of common and scientific names, 151–53. *See also* moas; New Zealand birds; parrots; raven, common
black market, 123, 126–27
blood poisoning in sheep, 133–34
bounties, 35–36, 128, 182n. 24
breeding. *See* reproduction
brood size, 82–83; for kakas, 115–16
Buller, Sir Walter, 30–31, 36, 39–40, 182–83n. 28

Canterbury District, 29, 34–35, 51, 145
carnivory. *See* diet: predation
carrion. *See under* diet
censuses, 50–56, 109–10. *See also* population, kea
Christchurch, 3, 29, 92, 124, 136
CITES (Convention in International Trade in Endangered Species), 129
classification: of keas and kakas, 30, 178–79n. 25; of moas, 43–45, 175n. 3
Clostridium bacteria, 133–34
clutch size, 82–83; for kakas, 115–16
colonists, 31–32
color patterns, kaka: body feathers, 27, 30, 102; face, 104–5; underwings, 111, 112. *See also* age determination: in kakas
color patterns, kea, *frontispiece*; body feathers, 14, 27, 102; face,
51–52; underwings, 1, 91. *See also* age determination: in keas; physiology: sexual maturation
comparative ethology, methods of, 101–2, 194n. 1
competition, ecological: with kaka, 17–18, 24, 105, 147, 180n. 33, 181–82n. 21; with other species, 17–18, 42
conservation. *See* New Zealand Department of Conservation
Cook, Captain James, 29–30, 42
courtship. *See under* sexual behavior
Craigieburn Forest Park, 29, 143, 177n. 14
crouch display: age and sex differences, 68, 93–94, *160–61*, 188n. 29; ambiguity in, 68–69; behavior pattern, 62, 67–68; as solicitation, 66–68, 76; as submissive behavior, 68, 92–94, 126
culmen. *See* morphology: bill
cultural transmission. *See* social learning
curiosity. *See* object play

Darwin, Charles, 30, 101
demolition, 39–40, 58, 78–79, 85–86, *87*, 92, 96, 134–38. *See also* object play: age differences; object play: and demolition
density, population. *See* population, kaka; population, kea
Department of Conservation. *See* New Zealand Department of Conservation
destructiveness. *See* demolition
development: contrast between kea and kaka, 111; fledglings, 84–89; juveniles, 89–92, 191n. 14; nestlings, 82, *83*; sex differences

in, 74, 88–89, 94, 99–100; sub
adults, 93–95; time course of,
51–52, 82–83, 94, 191n. 2
Diamond, Jared, 2
diet: breadth of, *17*, 18–19, 42; carrion, 4, 12, 14, 16, 32–34, 40, 45,
106, 141, 146–48; fondness for
fat, 16, 50, 89, 132; of juveniles,
89; of kaka, 11, 105–6, 114–15,
195n. 6; predation, *17*, 33–40,
177n. 15; reliance on southern
beech, 14–15, 18, 58, 150; seasonal
variability, 15–16, 177n. 14; sex
differences in, 58
dimorphism, sexual, 52–53, 55
dinosaurs, age of, 8, 19–20
discovery of kea and kaka, 29–31
dispersal, 92, 94
display. *See* crouch display; dominance: stare-down display;
feather postures; hunch display;
pout display; social play; winghold display
distribution, kaka, 18, 105, 107,
195n. 5. *See also* population, kaka
distribution, kea: current, 141–45,
144; extension of range, 31,
33–34, 181–82n. 21; original, 2,
31, 179n. 29, 180n. 33; overlap
with kaka distribution, 105,
195n. 5. *See also* population, kea
dominance: and aggression, 59; determination of, *162–63;* and
feather postures, 60–62, *61;* in
foraging, 59–60; and hunching,
92, *173–74;* in object play, 190n.
40; in reproduction, 119–20; sex
and age differences, 60, *162–63,*
187n. 18; stare-down display, 60,
110; transitivity of, 60, *162–63,*
187n. 18

drinking, 185–86n. 13
dump, refuse: age and sex differences in use of, 57–58, 64, 80–81,
187n. 15; attraction to, 33, 46;
foraging techniques in, 57–58,
84–85; hazards of, 85, 138–39;
management of, 49, 139–40;
number of keas using, 55–56;
preferred foods in, 58, 89; study
site description, 46–49

ethogram: of allopreening, 65–66;
of aggressive behavior, 59–62; of
crouch display, 67–68; of foraging behavior, 57–58, 105–6,
185–86n. 13, 186n. 14; of hunch
display, 89–91; of play behavior,
69–74, *164–65;* of pout display,
111–12, *113;* of wing-hold display, 86–88. *See also* feather postures
Europeans: explorers, 29–30,
182–83n. 28; forest destruction
by, 40–41, 107, 183n. 38; introduction of alien species by, 41–42,
107, 184n. 43; settlers, 31–32; and
sheep ranching, 32–36
evolution of kea and kaka, 21–22,
23, 179n. 29
explorers. *See under* Europeans
expressions, facial. *See* feather postures
extinction: mass, 21, 41–42; of New
Zealand birds, 18, 42, *43–45,*
184n. 43. *See also* parrots: endangered and extinct; parrots: Norfolk Island kaka

facilitation, social. *See* social facilitation
Fagen, Robert, 99

feather postures, 60–62, *61*
feeding, regurgitant: conflict with social play, 74–75; of kaka females by males, 111–12, *113;* of kea females by males, 82, 66, 68–69; role of male and female parent, 67, 82; solicitation by fledglings, 67–68, *67;* time course, 67, 82–83, 188n. 28
Fiordland National Park, 28, *29,* 105, 125, 145–46, 176n. 5, 177n. 14
flexibility: in choice of food, 4, 121–22; evolution, 24, 122, 147–50, 200n. 38; limitations, 4, 150; maintained by play, 99–100. *See also* open programs
flocking, 86, 92
food preferences. *See* diet
food resources: contrast between kea and kaka, 114–15, 121–22; limitations, 114, 117–18; unpredictability, 121. *See also* diet
foraging: age and sex differences, 58, *157–58;* in fledglings, 84–85; in juveniles, 89–92; in kakas, 105–6; in subadults, 93–94; techniques, 57–58, 84, 89, 94, *157–58,* 185–86n. 13. *See also* demolition; diet; stealing
forest: destruction, 40–42, 183n. 38; distribution, 8, 18–19, 23–24, 48. *See also* southern beech
fossil record, 14, 23–24, 102, 179n. 29, 181–82n. 21
Fox Glacier. *See* Westland National Park

Geist, Valerius, 99
generalist, ecological. *See* diet: breadth of

Gondwanaland, 19–20
Gould, John, 30

Haast, Sir Julius, 39
hazards: dumps, 85, 138–39; poisoning, 85, 140–41, 198n. 21; shooting, 35–36, 128–29; smuggling, 123–25, 127
heat tolerance, 56, 64–65
Hector, Sir James, 31
holocaust, biological, 7
humans in New Zealand. *See* Europeans; Maori
hunch display: absence of in kakas, 111; age and sex differences in, 90, 93, *160–61;* context and development, 90–92; and dominance, 92, *173–74;* form of, 89–90, *90;* lack of effect in subadults, 93
Huxley, T. H., 30

imitation. *See* social learning
introduced species: cat, 41, 62, 108, 116; deer, red, 41, 146, 183n. 40; dog, 41; *kiore* (Polynesian rat), 41–42; pigs, goats, and cattle, 35, 108; plants, 41, 150; possum, brush-tailed, 41, 108, 140–41, 183n. 40, 198n. 23; rabbit, European, 18, 34–35, 41, 184n. 44; rat, Norway or Black, 41, 108, 116, 142, 193n. 23, 196n. 23; stoat, 41–42, 62, 116. *See also* predation: by introduced species

Jackson, J. R., 46–47, 94, 111, 115, 117–20, 146

Kahurangi National Park, *29,* 145
Kapiti Island, *3,* 105–10, 116–17, 121

Keulemans, J. G., 36
kleptoparasitism. *See* stealing

Latham, John, 30
learning: of foraging techniques by fledglings, 84–85; role in kea society, 84, 95–100. *See also* flexibility; open programs; social learning
Lorenz, Konrad, 194n. 1

mammals. *See* bats; introduced species
Mantell, Walter, 30
Maori: forest destruction by, 40–42, 183n. 38; hunting moas, 40, 42; relationship to kaka and kea, 27–28, 180nn. 4–5; settlement, 7, 26–27, 40–42, 107
Marlborough District, 29, 32
Marriner, George, 36, 145, 181n. 21, 182n. 26, 189n. 31
mating, 66, 68
Mayr, Ernst, 148
McLean, Malcolm, 108
methods: at Arthur's Pass National Park, 47–49; general techniques, 50–56, 109–10; on Kapiti Island, 108–10
moas: classification, *43–45,* 175n. 3; competition with keas, 17–18; ecology and size of, 8–10, 12, *13,* 175n. 3, 176n. 4, 177–78n. 18; extinction of, 18, 28, 41–42; origin of, 2, 20, 178n. 21, 179n. 28; scavenging of, by keas, 12, 14
Moorhouse, Ron, 109, 115, 117, 196n. 25
morphology: bill, 53, *54,* 57, 102, 106, 185–86n. 13; crown and nape, 60–62, 66–67; differences between kea and kaka, 102–5, *104,* 106, 195n. 3; feet and legs, 57, 102–4, 186n. 13; sex differences in, 52–53, *55;* specializations for foraging, 57, 106, 185–86n. 13; tongue, 57, 186n. 13; weight, 53, 102. *See also* color patterns, kaka; color patterns, kea
mortality: comparison between kea and kaka, 116–17, 196n. 25; juvenile, 116–18, 120. *See also* hazards; predation
Mount Aspiring National Park, 29, 143–45
Mount Cook National Park, 29, 81, 142–43

names, common and scientific, list of, 151–55
Nelson District, 29, 105, 141–42, 181–82n. 21. *See also* Nelson Lakes National Park
Nelson Lakes National Park, 29, 143
nesting. *See* reproduction
Nestor, 21, 30, 102, 179n. 25. *See also* parrots: Norfolk Island kaka; proto-kaka
New Zealand: changes in land mass, 19–22, *23;* geography of archipelago, 2, *3. See also* North Island; South Island; Stewart Island
New Zealand birds, native: adzebill, 12, 18; bellbird, 11, 107; cuckoo, long-tailed, 107; eagle, Haast's, 12, *13,* 18, 42, 176n. 10; falcon, New Zealand, 12, 63–64, 188n. 21; fantail, 11; goshawk, New Zealand, 12, 18, 42; gull, black-backed, 12, 17, 62, 133; harrier, Australasian, 12, 17, 131,

New Zealand birds, native (*continued*)
133, 184n. 44; huia, 11; kiwi, 2, 11, 107; kokako, 11; morepork, 12; owl, laughing, 12; owlet-nightjar, 11; pigeon, New Zealand, 11–12, 27, 63, 107; piopio, 11; pipit, 62; pukeko, 63; raven, New Zealand, 12, 14, 17–18, 42; rifleman, 62; robin, New Zealand, 11, 62, 107; rock wren, 62; saddleback, 11, 107; silvereye, 62; snipe-rail, 42; stitchbird, 11, 107; takahe, 10, 17–18, 107, 176n. 5; tui, 11, 63, 107; weka, 11–12, 177n. 15; whitehead, 11; wren, flightless, 2, 10, 42, 176n. 6, 184n. 43; yellowhead, 11. *See also* extinction: of New Zealand birds; moas; parrots: kakapo; parrots: New Zealand parakeets (kakariki)

New Zealand Department of Conservation, 46, 49, 51, 108, 124, 130–31, 136–39, 146

North Island: geography, 2, 3; keas on, 150, 179n. 29, 181–82n. 21; Maori, 27; sheep raising, 32. *See also* distribution, kaka; Kapiti Island

Nothofagus. *See* southern beech

object play: age differences, 78–79, 85–86, 93, *157–58;* and demolition, 39–40, 78–79, 85, *87,* 96; and dominance, 190n. 40; facilitation of, 78–79; and harassment of animals, 62, 147–48, 200n. 34; in kakas, 108–9, 114; in other bird species, 188–89n. 31; persistence in, 39–40, 78–79, 135, 190n. 42;

and properties of object, 76–78, 190n. 41; in social context, 77, 79 observational learning. *See* social learning
O'Donnell, Colin, 110
ontogeny. *See* development
open programs, 4, 95–96, 148. *See also* flexibility
Otago District, 29, 31, 36, 145

parental care, 67–69, 82–83, 86. *See also* feeding, regurgitant
parrots (other than kea and kaka): Australo-Papuan, evolution, 21, 178–79n. 25; endangered and extinct, 129, 141–42, 198n. 26; illegal trade in, 124, 127, 129; kakapo, 2, 10, 12, 17–18, 28, 42, 141, 176n. 5, 179n. 25, 183n. 28; New Zealand parakeets (kakariki), 11, 63, 107, 141–42; Norfolk Island kaka, 21, 142; play, 69, 188–89n. 31; social learning, 149. *See also* proto-kaka
peripatus, 19–20
Philippines, 42, 184n. 43
physiology: caloric requirements, 117, 196n. 26; sexual maturation, 94–95, 192–93n. 18; thermal tolerance, 56, 64–65
piracy. *See* stealing
play: comparative study, 197n. 35; definition of, 189n. 32; description, in keas, 4, 69–81, *164–165,* 189–90n. 33; developmental changes, 99–100, *160–61;* and flexibility, 99, 122; in other birds, 188–89n. 31; signals, 72–73. *See also* object play; social play
plumage. *See* color patterns, kaka; color patterns, kea

Polynesian. *See* Maori
population, kaka: decline since colonization, 18, 34, 116, 146; on Kapiti Island, 107, 110; original abundance, 18, 116. *See also* distribution, kaka
population, kea: at Arthur's Pass dump, 55–56, 140, *156;* current levels, 141–43; estimation methods, 199–200n. 29; fluctuations, 23–24, 28, 30, 33–34, 42, 45, 116, 118, 145–47. *See also* distribution, kea
Potts, T. H., 37
pout display, 111–12, *113*
precipitation: in Arthur's Pass National Park, 48, 117–18, 185n. 6; on Kapiti Island, 106–7, 121
predation: by extinct raptors, 12–14, *13*, 42, 176n. 10; by falcons, 12, 63–64; by introduced species, 41–42, 116, 196n. 23
predation by keas. *See under* diet; sheep
protection, legal, 127–29; penalties, 125
proto-kaka, 21–22, *23*, 112

ranching. *See* sheep: ranching
range, geographic. *See* distribution, kaka; distribution, kea
raven, common: dominance displays, 62; maturation, 192–93n. 18, 194n. 1; and open programs, 96; play, 69
reintroduction, problems with, 125–27
reproduction: incubation and nesting success, 82–83, 115–16; in kakas, 115–18, 120; nest site limitations, 118–19; social constraints, 119–20, 196n. 31; timing of first, 83, 94–95, 120–21, 191n. 2, 196–97n. 32
reptiles, New Zealand: skinks and geckos, 10, 19–20; tuatara, 10–11, 14, 19–20, 128, 176n. 6
Ritzmeier, Michaela, 78
Routeburn Basin, *29,* 146, 177n. 14

Sainte Croix de Belligny, 30
scavenging. *See* diet: carrion
scientific names, list of, 151–55
settlers. *See under* Europeans; Maori
sex ratio, 58, 110
sexes, identification of. *See* morphology: sex differences in
sexual behavior: courtship, 69, 71–72, 74, 93, *160–61, 170–71;* in kakas, 111–12, *113;* mating, 66, 68; pair-bond maintenance, 69, 74
sheep: blood poisoning, 133–34; predation by keas, 4, 32–40, 130–34, 147–48, 182n. 26, 182–83n. 28, 200n. 34; predation by other birds, 106, 132–33; ranching, 31–36, 108, 128, 134, 145
smuggling, 123–25, 127
snails, land, 10, *17,* 19
social behavior. *See* aggression; allopreening; crouch display; dominance; feather postures; feeding, regurgitant; hunch display; object play; pout display; sexual behavior; social play; submissive behavior; vocalization; winghold display
social facilitation: definition, 96–97, 193n. 21; developmental changes, 98–100; in foraging, 86, 96–97, 149; limitations on, 97–98, 193–94n. 24; in object play, 78–79; in social play, 73–74, 100, *166–69*

social learning, 97, 149, 193nn. 22–23, 200n. 38. *See also* social facilitation
social play: absence of in kakas, 114; age differences, 74, 93, 99–100, *160–61, 172;* conflicts over, 74–76; facilitation, 73–74, *166–69;* sex differences, 71–72, 74, 93, 100, *160–61, 168–71;* toss play, 71–72, 74, 93, *160–61, 170–71;* tussle play, 69–71, *71–72,* 73–74, *166–69, 172,* 190n. 37
Southern Alps, 2, *3,* 8, 21–22, *29,* 31, 48
southern beech: distribution, 8, 19, 23–24, 145; mast-seeding, 15, 177n. 13; structure of forest, 14–15, 49, 183n. 40; use by keas, 14–16, 18, 58, 105–6, 114, 118, 121, 145, 150
South Island: birds, 9–10, 176n. 5, 177n. 18, 188n. 21; changes in, 23–24, 147; European settlement, 30–32; geography, 2, *3, 29,* 46, 48; Maori, 26–28; sheep raising, 32, 130. *See also* distribution, kaka; distribution, kea
Southland District, *29,* 34
stealing, 58, 77, 94, *157–58,* 192n. 16
Stewart Island, 2, *3*
study sites, *29,* 47–49, 108–10. *See also* Arthur's Pass National Park; Kapiti Island
submissive behavior: contrasted with appeasement, 91–92; displayed in feather postures, 62. *See also* crouch display
survival. *See* mortality

Tasman Mountains, *29,* 145
Tasman Sea, *3,* 20
taxonomy. *See* classification
teaching. *See* social learning
temperature: in Arthur's Pass National Park, 48–49; influence on foraging, 56, 64–65, 117; on Kapiti Island, 107, 117
Temple, Philip, 134
theft. *See* stealing
toss play. *See under* social play
trade, pet. *See* parrots: illegal trade in
tuatara. *See under* reptiles
tussle play. *See under* social play

vigilance: caution in approaching dump, 56–57; while foraging, 63–64
vocalization: in aggressive interactions, 80, 87, 90; alarm calls, 63; bleat-trill and flight call, 27, 56, 80; in play, 70; repertoire, contrast between kea and kaka, 112–13

Wallace, Alfred Russel, 182–83n. 28
Westland District, *29,* 105. *See also* Westland National Park
Westland National Park, *29,* 47, 143, 185n. 1
wetas, 10, 19, 38, 176n. 6
Wilson, Kerry-Jayne, 51, 142–43
Wilson, Peter, 110
wing-hold display, 86–90, *88,* 94

Compositor:	Impressions Book and Journal Services, Inc.
Text:	11/15 Granjon
Display:	Weiss
Printer:	Malloy Lithographing, Inc.
Binder:	John H. Dekker & Sons